42 Advances in Biochemical Engineering/ Biotechnology

Managing Editor: A. Fiechter

Bioprocesses and Applied Enzymology

Guest Editor: J. Reiser

With contributions by
R. Blaszczyk, W. H. J. Boesten, Q. B. Broxterman,
I. Endo, K.-E. L. Eriksson, H. F. M. Hermes,
H. Inaba, J. Kamphuis, N. Kosaric, E. M. Meijer,
T. Nagamune, H. Nakajima, T. Ohshima,
K. Ramasubramanyan, H. E. Schoemaker, K. Soda,
S. Tachikawa, A. Tanaka, J. A. M. van Balken,
L. Vallander, K. Venkatasubramanian

With 83 Figures and 66 Tables

Springer-Verlag
Berlin Heidelberg GmbH

ISBN 978-3-662-15075-7   ISBN 978-3-540-47150-9 (eBook)
DOI 10.1007/978-3-540-47150-9

© Springer-Verlag Berlin Heidelberg 1990
Originally published by Springer-Verlag Berlin Heidelberg New York in 1990
Softcover reprint of the hardcover 1st edition 1990

Library of Congress Catalog Coard Number 72-152360

Typesetting: Th. Müntzer, Bad Langensalza;

2152/3020-543210 — Printed of acid-free paper

Editorial

This volume of Advances in Biochemical Engineering/Biotechnology as well as the following one are dedicated to Professor Armin Fiechter on the occasion of his 65th birthday. Contributions were solicited from Armin's colleagues but were limited to subjects which related to the biotechnology scope of this series.

One might judge a successful scientist by several criteria and Armin fullfills them all. One criterion of success is one's record of scientific achievement. Armin's work on yeast physiology, process development and instrumentation has been very fruitful and the necessity of adopting new technology or applying concepts from other disciplines has never hindered progress in his laboratory. One very important measure of achievement and the one which is likely to be the most important in both professional and human terms, is the training of successful students. Armin is one of the most eminent promoters of biotechnology in Switzerland and abroad and he was instrumental in establishing the Institute for Biotechnology af the ETH in 1982 of which he is still director. By pioneering the combination of fundamental and applied aspects he has greatly influenced the advancement of biotechnology in teaching and research in Switzerland.

At the age of 65, the pace of Armin's activities shows no sign of slackening. At the latest count there are some 20 doctoral students working under his supervision. He is currently the editor-in-chief of the Journal of Biotechnology and of this series, both of which were co-founded by him and have gained momentum under his influence.

This special issue is dedicated to Armin Fiechter with admiration.

Zürich, July 1990   Jakob Reiser

Armin Fiechter

Table of Contents

A Tubular Bioreactor for High Density Cultivation of Microorganisms

Isao Endo[1], Teruyuki Nagamune[1], Susumu Tachikawa[2], Hideki Inaba[2]

[1] The Institute of Physical and Chemical Research, Wako-shi, Saitama, Japan
[2] Sumitomo Heavy Industries, Ltd., Hiratuka-shi, Kanagawa, Japan

By simulating the functions of the animal intestine, the authors have developed a novel tubular bioreactor (TBR) which is capable of containing both the reaction and separation of products in a single system. This reactor consisted of inorganic ultra filtration membrane modules in the primary part of the system, a heat exchanger and a recycling pump.

   The operation characteristics of the TBR were studied by cultivating *Lactobacillus casei* at a laboratory scale. The cell density obtained was up to 10 times higher than the density obtained by using the conventional jar fermentor. Furthermore, 40 g l^{-1} of cell mass was obtained in only 14 hours with 20 l of fresh medium when the dilution rate was increased according to the cellular growth. Afterwards, the cultivation time and the volume of fresh medium were reduced to 44% and 74%, respectively, of the values in the cultivation operation at constant dilution rate.

Advances in Biochemical Engineering/
Biotechnology, Vol. 42
Managing Editor: A. Fiechter
© Springer-Verlag Berlin Heidelberg 1990

1 Introduction

First of all, we wish to express our heartfelt congratulations to Professor Armin Fiechter on the occasion of his 65th birthday. Recognizing the importance of bioreactors in the field of bioprocesses, Professor Fiechter has for many years been at the forefront in developing bioreactors. The outcome of this research is condensed in his recent review paper of "Physical and Chemical Parameters of Microbial Growth" [1].

The separation-capable bioreactor being currently developed integrates both the reaction process and separation in a single system and has attracted the attention of many biochemical engineers around the world [2, 3]. Due to the characteristics of the reactor, high density cultivation of the cells is possible overcoming the product inhibition which is inherent in the microbial growth reaction. The following three categories of reactors are currently being developed:
1) A bioreactor which is composed of a jar fermentor and separation unit [4], such as a centrifuge and a settler.
2) A bioreactor in which the biocatalysts are immobilized in its primary part [5].
3) A bioreactor which encloses and circulates the biocatalysts inside a membrane modules [2, 6, 7].

The former two types of reactors have been developed so far. However the last type of reactor has never been achieved because of difficulties with the membrane modules and their effective operation. The authors have succeeded with this type of reactor by producing a TBR [8]. By incorporating, as the primary part of the reactor, an inorganic UF membrane which has advantages in sterilizability, in chemical properties and in filtration performance, we have made the TBR.

The aim of this paper is to introduce the design and operational characteristics of this TBR and how it is applied to the high density cultivation of *Lactobacillus casei*.

2 Design of the TBR

2.1 Membrane Module

There are various membrane materials which may be used for the TBR. The requirements for the membrane materials are as follows:
(1) Durability against steam sterilization; i.e. bioreactor elements are needed which are durable against repeated steam sterilization.
(2) Full recovery of filtration flux by cleaning. The decrease of a filtration flux caused by fouling with chemical substances present in the culture medium or with microorganisms affects the efficiency of the membrane bioreactor. This situation requires that the membrane flux should be recovered fully by cleaning.

(3) Sanitary safety. The membrane module must be designed properly and must be made of appropriate materials so that any contamination from it is avoided completely [9].

The current advances in membrane manufacturing technology are remarkable, though organic membranes still seem to have some problems concerning the above mentioned requirements [9]. On the other hand, inorganic membranes which satisfy these requirements have been developed. Taking these requirements into consideration, the authors have used an inorganic membrane module made of zirconium oxide for the design of the separation-capable bioreactor.

2.2 Constitution and Characteristics of the TBR

As indicated in Fig. 1 [3], a conventional membrane reactor consists of a jar fermentor and a membrane module like a hollow fiber type or a spiral type made of organic membranes. Generally speaking, the volume of the inorganic membrane module per unit area of filtration surface is larger than the one of the organic membrane module. Therefore, when this module is used as the separation

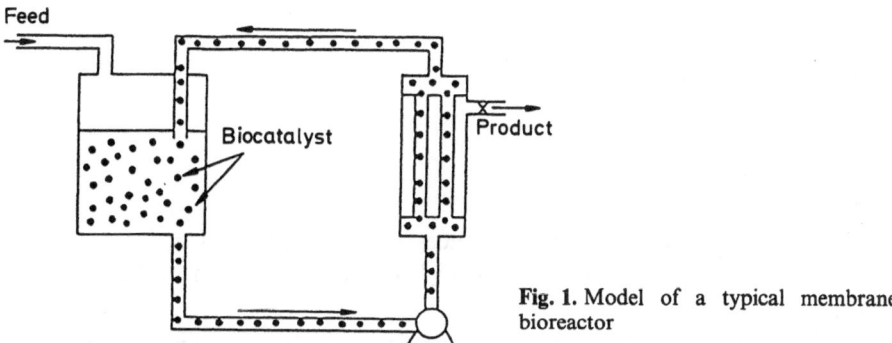

Feed

Biocatalyst

Product

Fig. 1. Model of a typical membrane bioreactor

unit, the working volume of the separator becomes significantly larger. Since the ratio of the fermentor volume to the separator volume decreases drastically proportionally to the increase of filtration surface area per unit volume of the fermentor, the fermentor loses its role as the site of reaction. Taking into consideration the characteristics of the inorganic membrane, we have developed a closed-loop tubular bioreactor which consists solely of an inorganic membrane module, a heat exchanger, a recirculating pump and pipes that connect the above components. The concept of the TBR is illustrated in Fig. 2. We can combine the TBR with a degassing device and an oxygen supply for aerobic cultivation use.

Since the TBR is a closed-loop tubular reactor, it has the following characteristics:

(1) The TBR is easy to scale up due to its simple configuration.

(2) As the TBR is operated under a pressure of 2 to 4 kg cm^{-2} for cross flow filtration, contamination can easily be avoided and the oxygen concentration and its effective use are high.

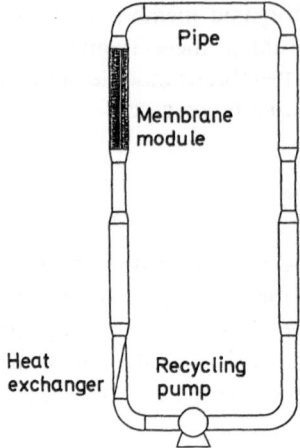

Fig. 2. Concept of a Tubular Bioreactor

3 Experimental Details

3.1 Equipment

In order to test the basic performances of the TBR, we set it up on a laboratory scale and applied it to the high density cultivation of *Lactobacillus casei*.

Figure 3 shows a flow diagram of the equipment. The total length of the closed-loop tube was 12 m and the reaction volume was 2.6 l. We used four UF membrane modules in which a tube with an internal diameter of 6 mm, an external diameter of 10 mm and a length of 1070 mm was introduced.

During circulation, the culture broth was filtered through the membrane tangentially and the product, lactic acid here, was removed. The rate of filtration was controlled automatically by a computer which was connected with a load cell.

3.2 Methods

3.2.1 Test Organism and Composition of the Basal Culture Medium:

Lactobacillus casei was used as a test organism in these experiments. The composition of the basal culture medium is shown in Table 1.

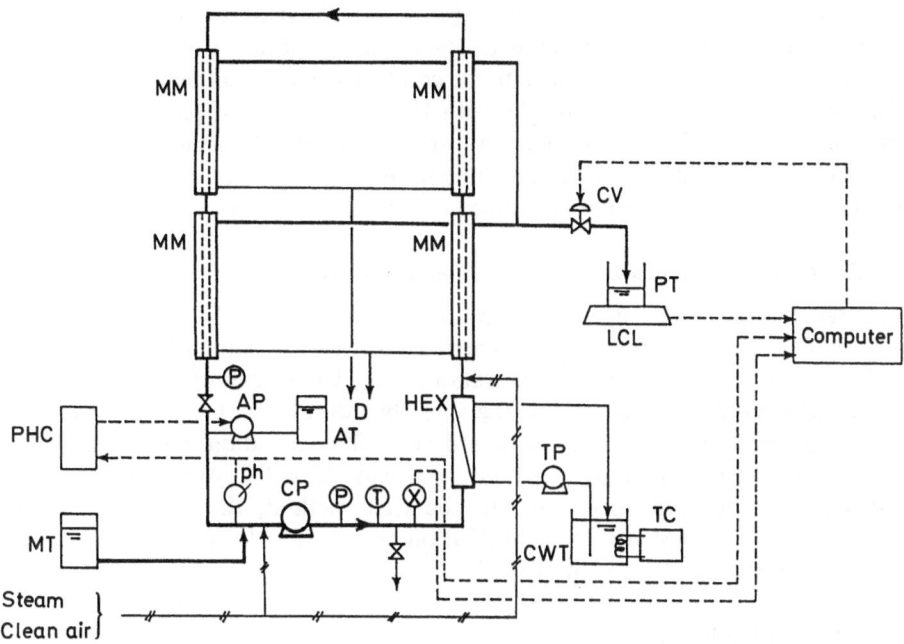

Fig. 3. Flow diagram of the TBR system; MM: membrane module CP: circulating pump HEX: heat exchanger, P: pressure gage, ph: pH sensor, T: thermometer, X: turbido sensor, pHC: pH controller, CV: control valve, LCL: load cell, TC: thermostat, AT: reservoir of alkali-sol., MT: reservoir of fresh medium, PT: reservoir of filtrate, AP: pump for alkali supplying, D: drain

Table 1. Composition of the basal medium

Substrate	Amount
Glucose	25 or 40 g l^{-1}
C S L*	100 or 160 g l^{-1}
K$_2$HPO$_4$	1 g l^{-1}
KH$_2$PO$_4$	1 g l^{-1}
MnSO$_4$	0.06 g l^{-1}

* Corn steep liquor

3.2.2 Experimental Conditions

3.2.2.1 Batch Culture

The TBR is usually operated under conditions appropriate for cross flow filtration, i.e. the flow rate on the surface of the membrane is 2 to 4 m s^{-1} and the average pressure operated is 2 to 4 bar. We studied the influences of operating conditions, such as the reactor characteristics derived from the closed-loop tubular type, the shear stress by the pump and other operating variables, on the microbial reaction. In order to elucidate the influence of the operational conditions on the microbial reaction, we conducted batch culture experiments in the TBR and compared the results with the ones in a mini jar fermentor (2 l working volume). The culture temperature, the initial pH value and the initial concentration of glucose in the medium were kept at 35 °C, 6.8 and 22 g l^{-1}, respectively. The pH was not controlled during batch culture.

3.2.2.2 Continuous Culture with Cross-Flow Filtration

The temperature and pH were kept at 35 °C $\pm$ 0.5 °C and 6.5 $\pm$ 0.2, respectively. Batch cultivation was carried out before continuous operation in order to increase the cell concentration. The batch culture was started at the initial glucose concentration of 40 g l^{-1}; when the glucose had been consumed, fresh medium was fed continuously and cross-flow filtration was started. The glucose concentration of the feed medium was 25 g l^{-1}. We conducted the following two feeding operations. In the first method, the feed rate of the fresh medium was kept at a constant dilution rate, 0.32 h^{-1}. In the second, it was raised according to the growth of the cells.

3.2.3 Measuring Methods

The culture broth was sampled manually every four hours in order to measure the dry weight of the cell mass, the glucose concentration and the lactic acid concentration. The concentrations of glucose and that of lactic acid were measured by using the new glucostat reagent and the F kit lactic acid reagent, respectively. In addition, the optical density (OD) was measured by an on-line turbido sensor (Laserturbidimeter, Model LT201, ASR in Japan).

4 Results and Discussion

4.1 Comparison of Batch Culture Experiments Obtained with the TBR with Those of the Jar Fermentor

Figure 4 shows the comparison of batch culture experiments obtained in the TBR with those of a jar fermentor. Table 2 represents the specific rate of cellular growth, that of substrate consumption, and that of lactic acid production during the exponential growth phase of the cells, which were calculated by data shown in Fig. 4. As indicated in Fig. 4 and Table 2, no significant difference in the results between two cultivation systems were observed. This means that the operating conditions and configuration of the TBR did not affect the reactions of the lactic acid bacteria.

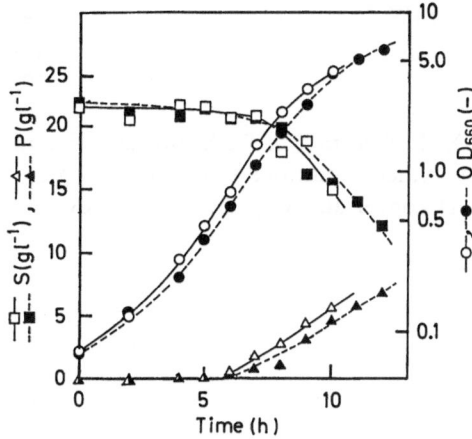

Fig. 4. Comparison of batch culture experiments obtained in the TBR with those obtained in a jar fermentor (JF)

Table 2. Values of the specific rates during the exponential growth phase of the cells in batch culture experiments

Reactor	μ (h^{-1})	v (h^{-1})	π (h^{-1})
TBR	0.533	2.25	1.96
JF*	0.462	2.20	1.74

* Jar fermentor

4.2 Cultivation of *L. casei* with Cross-Flow Filtration at a Constant Dilution Rate

The feed rate of fresh medium F, was matched with the rate of filtration in order to maintain the working volume, V, constant. As the mixing inside the TBR is perfect, the material balance concerning glucose, cell mass and lactic acid are given by equations (1) to (3).

$$
\begin{aligned}
dS/dt &= -vX + F/V \cdot (S_f - S) \\
&= -vX + D \cdot (S_f - S) \quad\quad (1)\\
dX/dt &= \mu X \quad\quad\quad\quad\quad\quad\quad\;\; (2)\\
dP/dt &= \pi X - D \cdot P \quad\quad\quad\; (3)
\end{aligned}
$$

where S_f is the glucose concentration in the feed of fresh medium, S is the glucose concentration in the reactor, X is cell mass concentration, P is the lactic acid concentration, D is the dilution rate, v is the specific glucose consumption rate, μ is the specific growth rate and π is the specific lactic acid production rate. Specific rates v, μ and π can be derived and calculated from the above equations.

$$v = -(dS/dt)/X + D \cdot (S_f - S)/X \tag{4}$$
$$\mu = (dX/dt)/X \tag{5}$$
$$\pi = (dP/dt)/X + DP/X . \tag{6}$$

Figure 5 shows the time course of S, X, P, feed rate of fresh medium, F, and amount of fresh medium supplied, $\int F dt$, under conditions where S_f is 25 g l^{-1} and D is 0.32 h^{-1}. The calculated results concerning v, μ and π are shown in in Fig. 6.

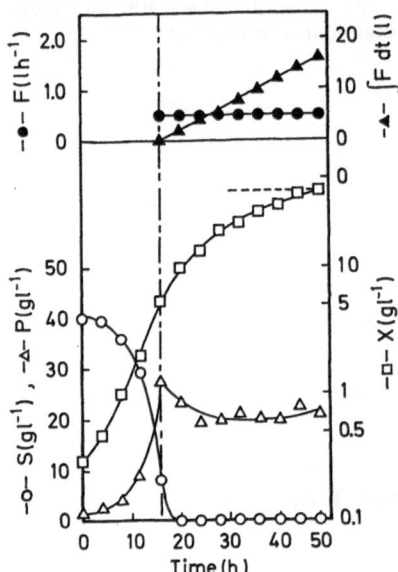

Fig. 5. Experimental results obtained by keeping the dilution rate constant

This figure shows that when S is decreased to nearly zero, the specific rates do not drop to zero immediately, but decrease gradually according to the lapse of time. When S and dS/dt in Eq. (4) are approximately zero, the specific glucose consumption rate, v, is obtained by the following equation;

$$v = D \cdot S_f/X . \tag{7}$$

Equation (7) indicates that the glucose consumption rate, v, is governed by the cell mass concentrations when $D \cdot S_f$ is constant. Since the specific growth rate, μ is influenced directly by the glucose consumption rate, μ is obviously also controlled by the cell mass concentration. This is the reason why the growth rate of the cell decreases when the cell mass concentration increases as shown in Fig. 5. From this consideration, it is suggested that, by increasing dilution rate according to the cellular growth, a high growth rate of the cell is to be expected.

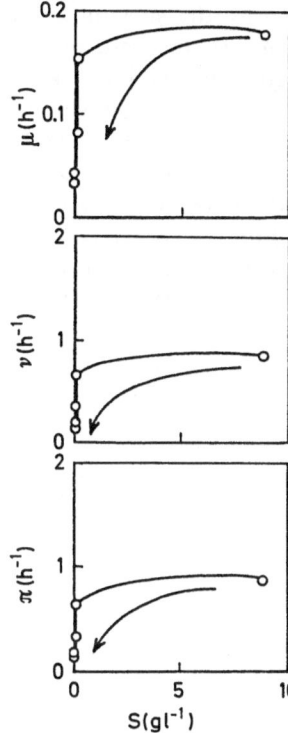

Fig. 6. Characteristic curves concerning to the specific growth rate of cells, μ, the rate of glucose consumption, ν, and the rate of lactic acid production, π vs the residual glucose concentration in the culture broth, S. The values were obtained under continuous cultivation conditions where the dilution rate was constant

4.3 Cultivation of *L. casei* with Cross-Flow Filtration by Increasing the Dilution Rate

On the basis of our knowledge mentioned above, we studied high efficiency and high cell density cultivation of *L. casei* with cross-flow filtration by increasing the dilution rate. From the stand point of efficient utilization of the feeding substrate, it is preferable that the glucose concentration in the reactor, S, should be kept at nearly zero. Under these conditions, the specific growth rate, μ, should be kept as high as possible to get a high cell mass concentration in a short time. Figure 6 shows that these requirements can be accomplished under the conditions that μ and ν are equal to 0.16 h^{-1} and 0.8 h^{-1}, respectively. Substituting this ν value and the value of S_f, 25 g l^{-1}, into Eq. (7), the dilution rate of fresh medium, D, is obtained by Eq. (8).

$$D = v \cdot X/S_f = 0.032X . \tag{8}$$

The cultivation of *L. casei* in the TBR was carried out controlling the dilution rate of fresh medium in proportion to the cell mass concentration. The cell mass concentration was estimated via a signal from the on-line turbido sensor. The experimental results obtained are shown in Fig. 7. The experiments shown in Fig. 5 and Fig. 7 are different only in the way the substrate was fed, but are identical in all other cultivation conditions. In the case of the experiment shown in Fig. 5, the dilution rate of the fresh medium was kept constant. Thirty two hours of cultivation time and 27 l of fresh medium were required so that the cell mass concentration reached 40 g l^{-1} (1.0×10^{11} viable cells ml^{-1}). On the other hand, the same concentration of cell mass, 40 g l^{-1} was achieved within 14 hours with 20 l of fresh medium by increasing the dilution rate of the fresh medium as it is shown in Fig. 7. The cultivation time and the volume of fresh medium used were reduced to 44% and 74%, respectively, of the values of the cultivation operation at a constant dilution rate.

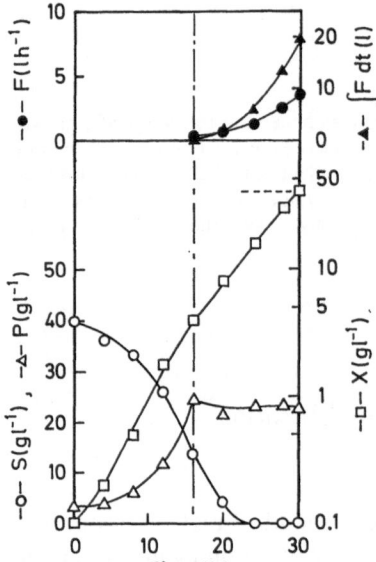

Fig. 7. Experimental results obtained by elevating the dilution rate according to the cellular growth

5 Conclusion

By using inorganic UF membrane modules in the primary part of the system, the authors developed a tubular bioreactor (TBR). Experiments for high density cultivation of *L. casei* were conducted in this TBR. We obtained cell density up to 10 times higher than obtained in the conventional jar fermentor. When the feed rate of fresh medium was elevated according to the cellular growth, the cultivation time and the volume of fresh medium used were reduced to 44% and 74%, respectively, of the values of the cultivation at a constant dilution rate.

6 Nomenclature

D   dilution rate (h^{-1})
F   feed rate of fresh medium $(g\ l^{-1})$
P   lactic acid concentration $(g\ l^{-1})$
S   glucose concentration in the reactor $(g\ l^{-1})$
S_f glucose concentration in the feed of fresh medium $(g\ l^{-1})$
t   time (h)
X   cell mass concentration $(g$ dry cell $l^{-1})$
μ specific growth rate (h^{-1})
ν specific glucose consumption rate (h^{-1})
π specific lactic acid production rate (h^{-1}).

7 References

1. Fiechter A (1984). Advances in Biochemical Engineering/Biotechnology. 30: 7
2. Nagamune T, Endo I, Fukuzumi M, Ohtake K (1987) Pro. of RIKEN Symp. on Separation. Capable Reactor, p 1
3. Cheryan M, Mehia MA (1986) Chemtec, Nov. p 677
4. Coutts MW: Australian Patent 216618
5. Kierstan M, Bucke C (1979) Biotechnol. Bioeng. 19: 387
6. Endo I, Inoue I: Japanese Patent 724683
7. Endo I, Inoue I, Tonooka Y.: Japanese Patent 937836
8. Tachikawa S, Inaba H, Seike Y, Nagamune T, Endo I (1988) Abstract of 8th International Biotech. Symp., Paris, p 165
9. Watanabe A (1987) Shokuhin Kogyo 1: 26

Large-scale Animal Cell Cultures: Design and Operational Considerations

K. Ramasubramanyan and K. Venkatasubramanian[1]
Dept. of Chemical and Biochemical Engineering, Rutgers, The State University of New Jersey, P.O. Box 909, Piscataway, NJ-08855 USA

The manufacture of biologicals, especially proteins, using large-scale culture of animal cells is becoming popular. There is a need for a rational approach to the design and scale-up of bioreactors for these applications. The ultimate requirement of any scale-up strategy should be to preserve the biological activity of these high-value molecules. With this as the central theme, the design and operation of animal cell processes has been discussed. Equal importance has been given to both the biological and the engineering aspects which need to be considered for a successful scale-up. An integrated systems approach has been stressed.

---

[1] Also with: H. J. Heinz Company, World Headquarters, Post Office Box 57, Pittsburgh, Pennsylvania 15230

Advances in Biochemical Engineering/
Biotechnology, Vol. 42
Managing Editor: A. Fiechter
© Springer-Verlag Berlin Heidelberg 1990

List of Abbreviations

D.O.:   Dissolved Oxygen
LDH: Lactate Dehydrogenase
ppm: parts per million
tpa: tissue plasminogen activator
MAb: Monoclonal Antibody

1 Introduction

The modern biotechnology revolution has led to the identification and subsequent production of a large number of new therapeutic proteins. Despite the early enthusiasm that these proteins could be produced in microorganisms such as bacteria and yeast — to take advantage of the commercial experience with these organisms and the relative ease of scale-up — a number of difficulties have restricted the use of these cells as host; These include:

• Many of the protein products of interest are glycosylated; this complex step is not easily achieved with these cells.
• The post translational modifications confer the crucial conformation and the desired biological activity on these proteins; again a step not inherently carried out by bacterial and yeast systems.
• The protein product is not usually secreted in these systems.

Therefore mammalian cells have increasingly become the vehicles of choice for the production of a large number of therapeutic proteins.

The use of hybridoma cells to produce many monoclonal antibodies (MAbs) — both for diagnostic and therapeutic purposes — has also spurred the study of large-scale animal cell cultures. While relatively small amounts of MAbs (of the order of few kg/year) can be efficiently and economically produced in vivo through the use of different animals (particularly mice) production of larger quantities often requires in vitro systems using mouse-mouse, mouse-rat, mouse-human and human-human hybridoma cell lines. In vitro production is the only meaningful method when human hybridomas are involved.

As the therapeutic protein products and MAbs move towards commercialization, a prodigious amount of work is being expended on studying the design and operation of mammalian/hybridoma cell culture reactor systems. The knowledge base for scaling-up these cultures, however, is rather limited. During the last few years, a number of bioreactor design concepts have been proposed and developed with varying degrees of success. These approaches range from suspension cell cultures — with and without an air lift provision to improve oxygen transfer — to attachment of the cells to a variety of matrices. However, only very few of these designs have the real potential for an economical large-scale operation.

In this paper we discusse several important factors that are critical to the successful design and operation of a large-scale animal cell bioreactor system and place the design problem within a practical commercial perspective. Although the focus of this paper is on mammalian/hybridoma cells, it is equally applicable to other emerging areas such as insect cell culture systems.

2 The Design Challenge

Mammalian cell products under discussion here are characterized by their realtively small volume requirements yet with high unit values. The large production costs associated with these proteins stem from several factors:

- These proteins are produced in relatively small concentrations (typically below 1 gl^{-1}).
- Mammalian cells are relatively slow growing and require complex nutrients and growth factors — whose functions are not clearly understood — in the growth medium; some of these may interfere with the subsequent purification steps.
- The difficulty in the isolation and purification of the desired protein i.e. the extraordinarily high degree of purity, — the absence of any contaminants such as residual DNA — and biological activity of the finished protein product required.
- Very stringent requirements mandated by the regulatory agencies such as the Food and Drug Administration (FDA) with respect to the mammalian/hybridoma cell lines, the engineering of the expression system, the production process and the process equipment.

A successful large-scale bioreactor system design must reflect all of these cost factors. The interactions between the production stage and the down-stream processing stage must be integrated in the design, and as such it is imprudent to address the bioreactor design without considering it as part of an overall production scheme. It should also be recognized that 'scale-up' in this context goes beyond the conventional increase in the system's size, i.e. the production volume requirements are relatively small — even the largest commercial bioreactor systems may be only of the order of hundreds to a few thousands of liters in size. Therefore space-time productivity per se is not the crucial scale-up criterion.

The scale-up challenge is to develop a production system which is reproducible, reliable, cost-effective and one which meets all the FDA requirements. From a regulatory standpoint, predictable operation on a batch-to-batch (or long term, if continuous) basis is perhaps the most critical factor. The predictability issue assumes even broader dimensions when one considers the fact that the production protocols and equipments must be specified very early on as part of the regulatory approval process.

Finally, the overwhelming factor which underscores all these issues is that any successfully scaled-up design must preserve the biological integrity of the protein molecule through all the production and downstream processing stages.

The following sections elaborate on each of the specific factors crucial to answering the design and scale-up challenge.

3 Cellular Aspects

3.1 Cell Density

In order to optimize production rates and product concentrations, one needs to maximize the cell density in the culture. An increased titer of the product, especially proteins, means a reduction in downstream processing costs. An increased cell-density does not mean an increased requirement of serum in the medium as shown by Glacken et al. [1]. This is true since serum components are found to trigger or catalyze cellular reactions and are not consumed to a significant extent. Hence we can afford to have a higher cell density without increasing the serum content of the medium.

The typical maximum cell density reported in a suspension cell culture is about 2 to 3×10^6 cells ml^{-1}. In the case of immobilized systems, cell densities upto 2 to 3×10^8 cells ml^{-1} of matrix have been reported [2, 3]. The distinction between the viable and non-viable fractions is not usually made in the published literature. Obviously the viable cell fraction is the one of interest with respect to cell density discussions.

One important advantage of high cell density systems is that it is possible to drastically reduce the costs associated with separating out cells during downstream processing.

A major problem associated with high density systems is scale-up. This usually is because of oxygen limitation, due to higher demand and/or mass transfer effects.

3.2 Cellular Productivity

The amount of product produced by each cell is a direct result of its intrinsic cellular machinery and the genetic manipulations performed to enhance it. The cell-specific productivity is dependent among other things on gene copy-number, promoter efficiency, rate of transcription, translation and processing, product inhibition and product secretion rate. The gene-level manipulations in genetically engineered animal cells to express a protein of interest is less complex, since the machinery required for post-translation modifications — e.g. glycosylation and folding — are present in the cell itself. This fact also renders procaryotic organisms less attractive for making recombinant proteins whose proper biological, therapeutic and immunogenic activity are crucial requirements.

The ultimate requirement of a bioreactor and process design is to make it genetically limiting. It is the duty of the engineer to preserve the high levels of cell-specific productivity. This means providing the cells with the optimal environment conducive for production. Automatic control of important physical variables — temperature, pH and D.O. concentration — thus become necessary and are standard features in almost all the commercial bioreactor packages. Providing the optimal microenvironment in the case of immobilized cultures with cell density comparable to the tissue density, becomes very difficult. In addition to the readily apparent nutrient limitations, it becomes necessary to consider the dynamic interactions between the local microenvironment and the bulk conditions and the effect of the microenvironment on cell health, productivity, genetic stability and the nature of the products produced.

Considering high costs of the products and the investment intensive nature of cell culture processes, it may be worth implementing a more sophisticated computer monitoring and control to optimize the cellular metabolism. A case in point is the production of interferon by FS-4 cells in a 7.5 liter culture [1] resulting in the reduction of lactate production and determination of the optimal time to induce cells to produce interferon.

3.3 Attachment Characteristics

Almost all the cell lines used in animal cell-culture processes can be classified as being able to grow freely in suspension or requiring an attachment surface for growth; this anchorage dependency (or otherwise) is a result of its origin — normal and abnormal tissues and organs. The process employed with a particular cell-line is usually based on its attachment characteristics, although sometimes there are overriding factors dictating the process, such as hardware availability and federal regulations. The usual practice until recently was to screen different candidate cell types to produce the product of interest, followed by optimization with respect to growth and production. But the impressive progress being made in genetic engineering of animal cells is facilitating the incorporation of the desired genetic apparatus into a pre-selected host cell line with good scale-up capability. In short developing a 'work horse' host cell line to produce a product of choice is becoming a reality [4, 5, 6, 7, 8].

Cells capable of growing in suspension do not pose many problems in process decisions; this is because the experinece that has been gained from microbial cells in suspension can be fruitfully employed for animal cell cultures, although the degree of complexity is higher.

Various methods are available to attach/immobilize cells to accommodate the attachment requirement of anchorage-dependent cells (discussed in detail in the section on bioreactor considerations). One interesting aspect of most normal (non-tumorigenic) cells is their preference for not growing in multiple layers because of contact-inhibition which necessitates an anchor with a high surface area to unit volume ratio [9]. Several such high surface microcarriers have been developed to address this requirement. However, it has also been shown that microcarriers and other similar matrices providing a conducive atmosphere for growth not only lead to very high cell densities but also promote those cells with "contact-inhibition" to grow in multiple layers [9]. The conditions of high cell density and space non-availability significantly reduce the growth rates of cells and consequently increases the product production rates [10]. The cultivation of cells in three-dimensional matrices in a tissue-like environment has been shown to improve the health [3] and the cell-cell and the cell-matrix interactions have been stated to enhance the cell specific productivity [11].

3.4 Cell Rigidity

The absence of a cell-wall makes animal cells very fragile. This is also the reason why the fermenters developed for microbial systems cannot be used for animal cell culture without modifications. The fragility — sensitivity to shear — imposes serious limitations on the mode and rate of aeration and agitation. Insect cell lines and some hybridoma cell lines have been reported to be especially shear sensitive [12]. Extensive research has gone into characterizing shear in animal cell culture bioreactors; quantifying the effect of shear on cell growth, cell damage and lysis; developing novel methods of aeration/agitation to minimize shear; and proposing new techniques such as immobilization/entrapment to decrease the exposure of cells to conditions in the bulk environment.

Even though shear effects is an important consideration in the design and scale-up of animal cell bioreactors, it can be said with confidence that it is no longer a "barrier" [1]!

3.5 Genetic Stability

Since most of the protein products of commercial importance produced by cell culture methods are a result of genetic engineering manipulations of the cell's genome, the stability of this gene is an important factor. Hence whenever possible, the use of immobilization/entrapment or a segregated mode of cell culture is a better choice since the space limitation will force a reduction in the rate of mitosis, which means that the chances of losing the genetic information are reduced [12].

4 Bioreactor Aspects

From the above discussion, it is obvious that the bioreactor is the heart of the whole process, the design and selection of which are dictated by biological, engineering and economic factors. This fact is outlined in Fig. 1, where the interactions of the various factors are shown.

4.1 The Cell System

A clear knowledge of the cell system under consideration is highly essential. Bioreactors are usually classified as those that employ suspension cells or those that employ anchorage dependent cells. Hence, the attachment characteristics of the cells form the basis for the selection of the bioreactor type. The origin of the cell-line (mammalian, insect, hybridoma) has a great say on its behaviour in culture as well as its attachment depency. The type of product of interest — proteins, viruses etc. — is crucial in bioreactor selection and operation.

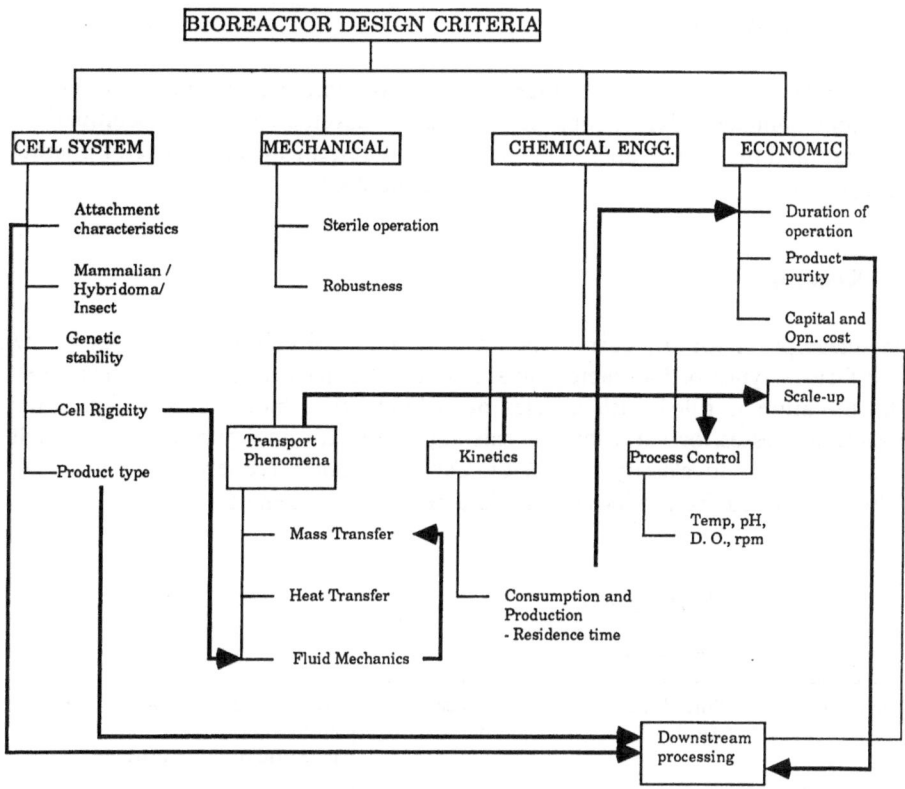

4.2 Chemical Engineering Aspects

All the major disciplines of chemical engineering play an important part in the design of a bioreactor. Mass transfer (convective and diffusional) exerts a great influence on substrate availability, product secretion and reactor scale-up potential. Proper analysis of the mass transfer rates in a bioreactor is the key to a successful design.

The fluid dynamics of agitation/aeration inside a bioreactor are critical because of the following two important considerations:

● It determines the availability of oxygen which is the least soluble, yet highly necessary nutrient and
● Since animal cells are very delicate, a high shear stress introduced into the bioreactor by agitation/aeration may lead to cell breakage.

The substrate consumption and product formation kinetics relate directly to the economics of the process by way of nutrient requirements and productivity of the process respectively.

Although the scale of operation is very much dependent on the demand for the product and the cellular productivity, the scale-up potential of a given reactor configuration is essential to its commercial success.

4.3 Mechanical Aspects

Structurally robust, easy-to-sterilize and easy-to-operate bioreactor systems are available commercially, thanks to the rapid development of biocompatible materials and the understanding of the aseptic requirements of bioreactor operation processes by mechanical and bioprocess engineers.

4.4 Reactor Types

One clear proof for the current boom in animal cell culture technology is the multifarious types of bioreactors available on the market for this purpose [13]. The literature is filled with articles describing novel bioreactors designed for specific processes, or variations of existing bioreactor configurations. Although it is impossible to classify all the bioreactor designs precisely and rigorously, Table 1 (adapted from Merten [12]) is a reasonable attempt.

Table 1. Bioreactor classification

	Low Density		High Density	
Reactor Types	Stirred tank, Airlift	STR with cell retention	Porous beads, Capsules (in packed/fluid-ized beds)	Hollow Fiber, Ceramic Cartridge
Typical Cell Density (ml^{-1})	$3-4 \times 10^6$	3×10^7	$1-3 \times 10^8$	1×10^8
Shear Protection of Cells	Yes (With good Design)	Yes (With good Design)	Yes	Yes
Downstream Processing	Difficult	Difficult	Easy	Very easy
Cell Monitoring and Control	Directly Possible	Directly Possible	Sometimes Directly Possible	Indirect — Glucose, Lactate, LDH etc.
Diffusion Limitation	No	No	Possible	Possible
Scale-up	Done	Done	Limited	Limited
Productivity	Low	High	Very High	Very High
Process Control	Easy	Easy	Slightly Difficult	Difficult
Cell-Type	Suspension	Suspension	Attachment Aggregate Suspension	Attachment Aggregate Suspension
Cell Growth Rate	High	Low	Very Low	Very Low
Products	Cellular Proteins, Vaccines, Proteins from Viral Infection	Cellular Proteins, Vaccines, Proteins from Viral Infection	Cellular Proteins, Vaccines	Cellular Proteins

The bioreactors available for batch or perfusion culture of suspension cells are in principle variants of classical stirred-tank fermenters used for microbial cell cultures. But in view of the delicate nature of the cell, the aeration/agitation schemes are designed to provide a gentle environment. Also because of their larger doubling time which make them more susceptible to contamination, the sterility requirement in terms of efficiency and duration are more stringent.

The perfusion systems available commercially belong to three different groups [12, 14]; The first is basically a stirred-tank system system provided with a means to retain the cells and characterized by microscopic homogeneity. One way is to use spin filters rotating at high speeds [15] and another is to use a second vessel with a filter which recycles the cells. The aeration/agitation methods are similar to those used for regular suspension cultures.

The second group of perfusion systems are those with macroscopic homogeneity. Cell containment is achieved through entrapment inside polymer gels or inside the pores of fibrous collagen matrices [2].

The third group includes those which are essentially non-homogeneous e.g. irregularities of ceramic cartridge surface and hollow fiber-systems.

One of the obvious differences between these systems in terms of process monitoring is the ability to estimate cell density and health on a regular basis. In the homogeneous systems it certainly is possible. In the second group of systems it is possible in some cases, for example using collagenase to free the cells immobilized in collagen microspheres [2]. In bioreactors that are non-homogeneous in nature, it is not possible to perform cell-counts regularly; instead, parameters like LDH concentration in the harvest medium, or oxygen or substrate consumption rate have to be used for process monitoring.

In the case of immobilized cell culture reactors, it is important to provide for the growth and maintenance of the cells, i.e. cells must be maintained in a viable state. Since the cells would leave the matrix — once the matrix is completely saturated — the use of a packed-bed reactor would be difficult as the free cells would tend to plug it. A fluidized-bed system would circumvent this problem. However, achieving an adequate density difference between the immobilized solid particle and the liquid medium — an essential requirement for operating a solid-liquid fluidized-bed reactor — is often difficult. Use of high density matrices is not necessarily an answer, as they are often not biocompatible in addition to having poor internal pore architecture to support cell growth. Weighting of porous beads is one approach to achieve both high cell density as well as adequate solid-liquid density differences [2, 3, 11]. Magnetically stabilizing a fluidized-bed is yet another approach.

4.5 Mode of Operation

The mode of bioreactor operation can be divided into three major classes — batch, fed batch and continuous. The relative merits and disadvantages readily obvious have been reported widely [1]. A few issues relevant to animal cell bioreactors are worth discussing. The choice of the mode of operation is dictated most of the time by the logistics of the production pattern. For example, the production of proteins by infecting the host cells with recombinant viruses [16] can only be carried out in a batch mode. In the case of products that are secondary metabolites or are formed by non-growth associated kinetics, the recommended mode of operation would be continuous; since the protein production rate is inversely related to the cell growth rate, operating a continuous culture at low growth rate conditions is economically beneficial.

An important determinant of large-scale cell-culture process economics is the cost of growing the cells for seeding the production bioreactors. In order to maximize the return on investing a given amount of time in growing cells, the production run should be carried out for as long as possible. This means that a continuous process is more economical when it is feasible. Also, the expenditures on quality control and product safety testing are lower for a continuous compared to a batch process, since the cumulative harvest gathered over a long period of time can be thought of as a single lot. In the case of products that degrade or become inactivated when left in contact with the culture, continuous operation is advantageous since the harvest can be purified more quickly. But one of the major disadvantages of a continuous/perfusion mode of operation is the loss of expensive medium components.

4.6 Scale-up

Industrial scale reactors with volumes ranging from 500 l to 3000 l have been used for vaccine production and 8000 l for interferon production [17]. But the strong theoretical and mechanistic basis necessary to give confidence in scale-up is not complete. Simple scale-up criteria such as power input per unit volume, mixing time, mass transfer, fluid velocity etc., which were found satisfactory for microbial systems are not sufficient for animal cell systems. The general philosophy behind scale-up and the use of dimensionless numbers as applied to suspension culture of animal cells is discussed by Bleim and Katinger [17, 18]. to the rate of molecular diffusion). In animal cell culture processes where scale-up For any given system the performance can be clearly illustrated by the use of dimensionless numbers since they readily show the relative importance of related effects (For example the Reynolds number compares the inertial forces to the viscous forces and the Sherwood number compares the rate of film mass transfer to the rate of molecular diffusion). In animal cell culture processes wehre scale-up is by empirical extrapolation, these dimensionless numbers can be very effective in transferring laboratory data to the design of the bioreactor.

Nutrient limitation — especially oxygen — and shear sensitivity have been the two major considerations in the published literature on animal cell bioreactors. Neither of them is sufficient by itself as a scale-up criterion since there are occasions when these two factors act against each other. Suspension cultures are better in terms of oxygen availability when compared to immobilized cell systems (where diffusional limitations may be severe) but are more susceptible to cell damage by shear, and vice versa. The reader is referred to the numerous papers that have appeared on the subject of bioreactor scale-up for specific approaches [17, 18, 19].

It should again be stressed here that increasing the bioreactor volume is not the issue; reproducing the physiochemical conditions is. Scale-up is dictated by auxiliary considerations which become crucial depending on the context:
— oxygen diffusivity through membrane and cell layers in hollow-fiber systems
— oxygen solubility in viscous cultures
— shear-sensitivity of cells in agitated/sparged systems
— cleaning-in-place and sterilization-in-place for certain bioreactor geometries.

Hence, a careful analysis of each of the component factors is highly essential. For large-scale systems, failure-analysis is a must; provision of redundancy is thus a common feature (e.g. membrane sterilization units kept in tandem to assure medium sterility).

5 Downstream Processing

One of the major steps that determines the profitability of a process is the purification/isolation of products from the harvest. As much as 80% of the manufacturing costs of the protein may be related to separation and purification. More important is the stringent regulatory requirement of safety of the final product and the process steps. Hence it is vital that the whole process and equipment be designed to minimize downstream processing costs and contamination by unacceptable substances.

Substances of potential biological hazard include viral or cellular DNA fragments, proteins produced by cells and viral particles. In products of clinical importance, proteins capable of inducing immune reactions in the host cells or cross reacting with other proteins are obviously unacceptable.

Factors affecting product safety/purification costs include:

5.1 The Product

If the product of interest is a protein, unsafe materials are eliminated by rigorous isolation/purification steps and confirming the product purity by analyzing the physicochemical properties. The extent of purification depends on the amount and the nature of the contaminants, the degree of purity necessary in the final product and the amount and nature of the product of interest.

In the case of vaccine production, since it is very difficult to eliminate the undesired products through a selective process, the best way to ensure biological safety and product potency is by closely controlling the process itself [20]. This includes thorough screening of the cells through about 50 generations and control of medium constituents and supplements during production.

5.2 Process Scheme

Almost invariably the first step in the downstream processing is the separation of cells from the culture liquor. Any kind of immobilized/entrapped mode of cell culture will drastically cut down this initial separation. Ball [21] has reviewed the various methods of cell clarification. Also the starting concentration of the product depends on the cell density, cell culture method and mode of operation for a given cell line.

5.3 Product Purity and Concentration

The final purity and concentration of the product desired depends on the specific application — therapeutic, diagnostic, immunopurification etc — and the product value. The purity of the product is a more crucial factor which calls for the absence of unneccessary substances or sets an upper limit on their concentration levels. From a consumer safety viewpoint the various National Regulatory Authorities are strict about the product putity. A protein contamination level of 100 ppm [22], nucleic acid contamination of 10–100 pg per dose [20, 23], absence of virus, microbes, mycoplasma and pyrogenic activity are all typically required.

5.4 Medium

The supplemental serum added to the medium for various purposes introduce extraneous proteins thereby complicating the purification process. The table comparing the recovery of recombinant tpa produced in serum-free and serum-containing medium by Cartwright [22] illustrates this point; Because of their animal origin and the inherent variability serum may introduce bacteria, viruses, yeasts or fungi into the medium. Hence, the serum should be thoroughly tested for sterility and also for the presence of endotoxin which is found to be difficult to get rid of during the purification processes.

The various purification, isolation and characterization methods have been reviewed [22]. Purification schemes for monoclonal antibodies have been discussed by Ostlund [24] and Rudge et al [25].

6 Cell Culture Medium

Historically one of the reasons microbial systems were chosen to express recombinant proteins in preference to animal cells is the need for complex and thus expensive culture media for animal cell cultures [13]. As shown by Griffiths [26] medium costs are typically as much as 40% of the total production costs and hence serious consideration of this factor is necessary before the process is finalized. Although it is highly tempting to buy a medium in the powder form and process it 'in house' the rigorous quality control requirement and a possible addition of antibiotics (which could amount to about 44% of the medium cost [26]) should be weighed. But with a streamlined lab, design and management techniques available (see the excellent review on this topic by Scheirer [27]) saving on medium costs could mean overall cost saving.

Serum could constitute about 17–84% of the medium costs [26]. Because of this and due to the fact that it could introduce unacceptable adventitious agents and variability between batches, omission of serum from the medium for production purposes should be seriously considered and investigated. As outlined earlier purification costs can be reduced significantly in the case of a serum/protein-free medium. The production of serum-free media has been forecast [28] as being a fast-growing market with 30–40% per year growth.

In the case of certain immobilized cultures, a switch from a serum-containing to a serum-free medium during continuous operation leads to increased cell-specific productivity as well as enhanced product purity [3]. Although the exact mechanism to explain this phenomenon is yet to be delineated, it appears that it is due to the unique microenvironment experienced by the cells in the immobilized state. Irrespective of the exact mechanism, this phenomenon is of great economic value both in terms of reduced medium cost and higher product purity.

7 Epilog

The need for structural, conformational and biological fidelity of protein products and their purity requirements challenges the bioreactor designer to an extraordinary degree with problems hitherto unencountered in conventional biological or chemical process design. The unique features of designing and The need for a different process engineering approach (as compared to scale-up of microbial systems) has been emphasized. The bioreactor design should be addressed as part of a well-wrought overall design scheme. A total systems approach is essential to ensure a successful design and reliable operation of the scaling-up of animal cell bioreactor systems have been reviewed in this paper. proach is essential to ensure a successful design and reliable operation of the biomanufacturing process. Reliability, predictability and regulatory compliance should be the major guiding principles.

8 Acknowledgements

The authors are grateful to Drs. Peter Runstadler, Nitya Ray and John Vournakis of Verax Corporation for their advice, counsel and encouragement in the preparation of this manuscript.

9 References

1. Glacken MW (1987) Large-scale production of mammalian cells and their products: Engineering principles and barriers to scale-up. In: Shuler ML, Weigand WA (eds.) Annals of the New York Academy of Sciences, Biochemical Engineering V. vol. 506 p. 355
2. Dean RC (1987) Large-scale culture of hybridoma and mammalian cells in fluidized bed bioreactors. In Shuler ML, Weigand WA (eds.) Annals of the New York Academy of Sciences, Biochemical Engineering V. vol. 506, p. 129

3. Vournakis JN, Hayman EG (1989) Cell microenvironment and the large-scale culture of recombinant Chinese Hamster ovary (CHO) cells: CHO cell series # 1. Attachment of cells to fibronectin in collagen microspheres and formation of an extracellular matrix. In: Verax Technical Bulletin
4. Wirth M (1987) Recombinant animal cell lines for production of glycoproteins. ESACT-OHOLO conference, p. 182
5. Silberklang M (1987) Foreign gene expression in rat pituitary GH3 cells. ESACT-OHOLO conference, p. 199
6. Wurm FM (1987) Use of transfected and amplified Drosophila heat shock promoter construction for inducible production of toxic mouse c-myc proteins in CHO cells. ESACT-OHOLO conference, p. 215
7. Weymouth L, and Barsoum J (1986) Genetic engineering in mammalian cells. In: Thilly W (ed) Butterworths, Boston, p. 9
8. Propst C, Hoppes H (1985) Soc. Ind. Microbio. News 35: 4
9. Tyo MA, Spier RE (1987) Enzyme Microb. Technol. 9: 514
10. Miller WM (1987) J. Cell. Physiol. 132: 524
11. Vournakis JN, Runstadler PW (1989) Biotechnology 7: 143
12. Merten O (1987) Tibtech (Aug): 230
13. Spier R (1988) Tibtech (Jan): 2
14. Arathoon WR, Birch JR (1986) Science 232: 1390
15. Van Wezel AL (1985) Dev. Biol. Standard. 60: 229
16. Inlow D, Harano D, Maiorella B. (1988) Bio/Technol. 12: 1406
17. Bliem R, Katinger H (1988) Tibtech (Aug): 190
18. Bliem R, Katinger H (1988) Tibtech (Sep): 224
19. Aunins JG (1086) Engineering developments in homogeneous culture of animal cells: Oxygenation of reactors and scale-up. In: Biotechn. Bioengg. Symp. No. 17, p. 699
20. Petricciani JP (1987) The liberation of animal cells: Psychology of changing attitudes. ESACT-OHOLO conference, p. 1
21. Ball GD (1985) Downstream Processing. In: Spierr RE, Griffiths JB (eds) Animal cell biotechnology, vol 2. Academic, London, p 87
22. Cartwright T (1987) Tibtech (Jan): 25
23. Liu DT (1985) Points to consider in the production and testing of new drugs and biologicals produced by recombinant DNA technologies. In: Report of the Recombinant DNA committee, National Center for Drugs and Biologicals, FDA
24. Ostlund C (1986) Tibtech (Nov): 288
25. Rudge J (1987) Continuous culture of murine hybridomas with integrated recovery of monoclonal antibodies. ESACT-OHOLO Conference, p 556
26. Griffiths B (1986) Tibtech (Oct): 268
27. Scheirer W (1987) Tibtech (Sep): 261
28. Jones SR (1987) Chemical Week 38

Microbial Aggregates in Anaerobic Wastewater Treatment

N. Kosaric
University of Western Ontario, Department of Chemical and Biochemical Engineering, London, Ontario, Canada, N6A 5B9

R. Blaszczyk
Lodz Technical University, Institute of Chemical Engineering, Bioengineering Section. 90-924 Lodz/Poland

The phenomenon aggregation of anaerobic bacteria gives an opportunity to speed up the digestion rate during methanogenesis. The aggregates are mainly composed of methanogenic bacteria which convert acetate and H_2/CO_2 into methane. Other bacteria are also included in the aggregates but their concentration is rather small. The aggregates may also be formed during acetogenesis or even hydrolysis but such aggregates are not stable and disrupt quickly when not fed. A two stage process seems to be suitable when high concentrated solid waste must be treated.

Special conditions are necessary to promote aggregate formation from methanogenic bacteria but aggregates once formed are stable without feeding even for a few years.

The structure, texture and activity of bacterial aggregates depend on several parameters: (1) — temperature and pH, (2) — wastewater composition and (3) — hydrodynamic conditions within the reactor. The common influence of all these parameters is still rather unknown but some recommendations may be given. Temperature and pH should be maintained in the range which is optimal for methanogenic bacteria e.g. a temperature between 32 and 50 °C and a value pH between 6.5 and 7.5. Wastewaters should contain soluble wastes and the specific loading rate should be around one $kgCOD(kgVSS)^{-1} d^{-1}$. The concentration of the elements influences aggregate composition and probably structure and texture. At high calcium concentration a change in the colour of the granules has been observed. Research is necessary to investigate the influence of other elements and organic toxicants on maintenance of the aggregates.

Hydrodynamic conditions seem to influence the stability of the granules over long time periods. At low liquid stream rates, aggregates may starve and lysis within the aggregates is

Advances in Biochemical Engineering/
Biotechnology, Vol. 42
Managing Editor: A. Fiechter
© Springer-Verlag Berlin Heidelberg 1990

possible which results in hollowing of aggregates and their floating. At high liquid stream rates the aggregates may be disrupted and washed out of the reactor as a flocculent sludge.

Methanogenic bacterial aggregates have been successfully applied in many full scale installations, especially for sugar beet, potato, pulp and paper mill, and other soluble wastes. The UASB reactors used for these treatments are simple in construction and handling which result in rather low total costs. A further and wider application of UASB reactors and methanogenic aggregates for various industrial wastewaters is expected.

1 Introduction

Wastewater treatment is becoming one of the most important problems of modern industry. Microbial digestion (aerobic and anaerobic) is mainly used to treat organic wastes.

Organic materials may be converted into methane and carbon dioxide, in the absence of exogenous electron acceptors such as oxygen, nitrate and sulfate, through a complex series of microbial interactions. In this process most of the chemical energy in the starting substrate is released as methane and may be recovered. In direct contrast, aerobic bacterial metabolism releases most of the original chemical energy from organic materials by oxidizing them to carbon dioxide and water; bacterial cells are also produced in large amounts.

The aerobic conversion of 1 kg COD (Chemical Oxygen Demand) requires 2 kWh of electricity (mixing and oxygen supply) and produces 0.5 kg of biomass (dry weight). Anaerobically, 1 kg COD gives rise to 0.5 m^3 biogas (equivalent to about 0.4 L of liquid fuel) and 0.1 kg of biomass which can be directly dewatered if required [1]. The anaerobic process not only results in an energy-rich product but also makes less cell material, and is consequently doubly useful in the degradation of biomass.

Tradionally, anaerobic digestion was utilized almost exclusively for the stabilization of sewage sludge. The process received little application in the treatment of organic industrial wastes due to several limitations, including the low achievable rates of performance, the inability to withstand hydraulic and organic shock loads, and the poor process control. These problems, all inherent to conventional digesters, were associated with difficulties in retaining biomass within the digester and with a very long retention time (up to 50 days) [2]. New rector designs and new methods of bacterial bed preparation have allowed this time to be shortened considerably — down to a few hours.

2 Anaerobic Conversion Processes

The decomposition of organic compounds by microbial cells involves two main metabolic processes, namely, dissimilation or biochemical oxidation of these organic compounds into energy and products, followed by assimilation of this energy by the cell which results in the growth and production of new cell material. Biochemical oxidation involves the reaction between two compounds in which protons and electrons are transferred from a reduced compound (the hydrogen electron donor) to a less reduced compound (the hydrogen electron acceptor). The electron acceptor can be either an inorganic compound or an organic compound. The inorganic compound may be oxygen (aerobic respiration), nitrate (nitrate reduction), sulphate (sulphate reduction) or carbon dioxide (methane formation). In bioprocesses, organic compounds act as the electron acceptors.

The anaerobic decomposition of organic matter is a complex process involving four successive stages realised by four types of anaerobic bacteria. In the first stage the hydrolysis of (mainly) suspended organics of high molecular weight into soluble simpler compounds (sugars, fatty acids, aminoacids, glycerol) is performed by hydrolytic bacteria. In the second stage the microbial processing of soluble organics into alcohols and carboxylic acids is performed by incompletely oxidizing bacteria. The third stage is the reduction of carboxylic acids and alcohols into acetate, hydrogen and carbon dioxide by acetogenic bacteria. Methane is produced in the last stage from hydrogen and carbon dioxide by hydrogenophilic methane bacteria or from acetate by acetophilic methane bacteria. A simplified scheme for anaerobic decomposition of organics is presented in Fig. 1.

Hydrolytic reactions generally limit the amount of methane being produced during anaerobic digestion of biomass, and also may be the rate limiting step

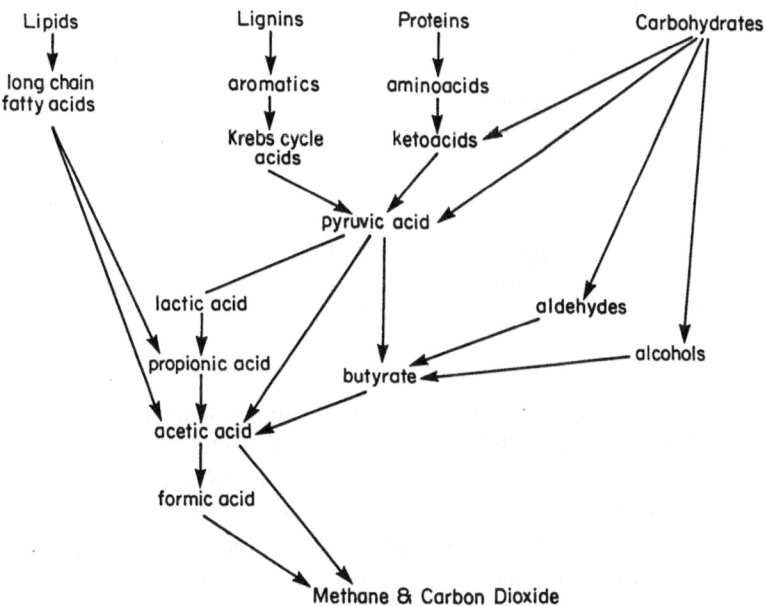

Fig. 1. Biochemical sequences for the breakdown of most important solids to methane and carbon dioxide — adapted from [2]

of the overall anaerobic microbial process [3]. Most of our knowledge of the anaerobic microbial breakdown of biomass polymers comes from studies of the microbial ecology of the rumen [4, 5]. Several microbial species which are reported to hydrolyse biopolymers anaerobically are presented in Table 1.

Table 1. Bacteria reported to hydrolyse biopolymers

Biopolymer	Bacteria	Ref.
Cellulose	*Ruminococcus flavefaciens*	[6]
	Ruminococcus albus	[7]
	Clostridium thermocellum	[8, 9, 10]
	Bacteroides succinogens	[11, 12, 13]
	Acetivibrio cellulotyticus	[14]
	Clostridium lochheadii	[7]
	Clostridium longisporum	[7]
	Clostridium cellobioporus	[15]
Hemicellulose	*Bacteroides ruminicola*	[16, 17]
	Bacteroides fibrisolvens	[16, 18]
	Bacteroides adolescentis	[19]
	Bacteroides eggerthii	[20]
Lignin	*Methanogenic* consortia	[21, 22, 23]
(after partially	*Pelobacter acidigallici*	[24]
solubilization)		
Pectin	*Lachnospira multiporus*	[25]
	Bacteroides fibrisolvens	[26]
	Bacteroides ruminicola	[27]
	Bacteroides succinogens	[28]
	Clostridium butyricum	[29]
	Clostridium multifermentans	[30]
Starch	*Streptococcus bovis*	[31]
	Bacteroides amylophilus	[32]
	Succimonas amylolytica	[33]
	Lactobacillus spp.	[34]
Lipids	*Anaevibrio lipolytica*	[35]
	Bacteroides fibrisolvens	[36]
	Butyrivibrio spp.	[37]
Protein	*Bacteroides amylophicus*	[38]
	Bacteroides ruminicola	[39, 40]

The major organic constituents of biomass are cellulose, hemicellulose and lignin, bound together in a lignocellulose matrix. Lignin is not only refractory to anaerobic decomposition but also affects degradation of cellulose, hemicellulose and other polymers. Extracellular enzymes must migrate into pore spaces (too small for bacteria) of lignocellulosic compounds to bring about hydrolysis. Factors affecting the slow rates of lignocellulose hydrolysis are still poorly understood [41]. Research is needed on hydrolytic bacteria and their enzymes, and the relationship between interpolymeric structure and their activities and kinetics. A full understanding of these interrelationships would allow the design of bioreactors with higher rate of degradation of polymeric compounds.

Incompletely oxidizing bacteria convert the soluble organic matter produced by hydrolytic bacteria into methanogenic substances (formate, acetate, H_2/CO_2, methanol) or non methanogenic substances e.g. propionic acid, butyric acid, alcohols and aldehydes, which are next converted by acetogenic bacteria into methanogenic substances. The schedule of reactions performed by incompletely oxidising and acetogenic bacteria are presented in Fig. 2.

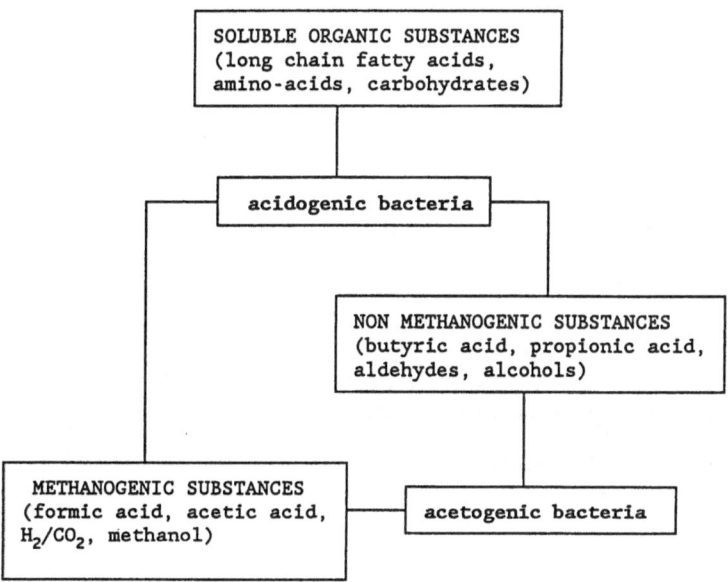

Fig. 2. Reactions performed by acid forming bacteria (acidogenic and acetogenic) — adapted from [42, 43, 44]

Acetate is the most important compound quantitatively produced in the bioprocessing of organic substrate by bacterial population with propionate production of secondary consequence [45]. There are a number of microorganisms which can digest soluble organic matter into acetic acid as *Lactobacillus, Escherichia, Staphylococcus, Micrococcus, Bacillus, Pseudomonas, Desulfovibrio, Selemonas, Veillonella, Sarcina, Streptococcus, Desulfobacter* and *Desulfomonas* [46].

The bioprocessing of the substrate to yield acetone, butanol, butyric acid and iso-propanol is mediated predominantly by bacteria of genera *Clostridium* and *Butyribacterium* [46]. The conversion of sugars to pyruvate via the Embden-Meyerhof-Parnas (EMP) pathway initiates the butyric acid bioprocessing.

Pyruvate breakdown is characteristic of the *Clostridia*. Pyruvate is decarboxylated to acetyl CoA, CO_2 and H_2 via a pyruvate-ferredoxin oxidoreductase; the lower ferredoxin redox potential preserves anaerobic conditions and prevents additional transportation of electrons [47]. The acetyl CoA maintained in equilibrium with acetyl phosphate by the enzyme phosphoacetyl transferase. The acetokinase enzyme then converts the acetyl phosphate to acetate with the production of one mole of ATP [15]. One mole of glucose produces therefore, 2 moles of acetate at the most, with 2 moles of CO_2 and H_2 evolved during pyruvate degradation.

Estimated free energy changes of selected biological reactions performed by incompletely oxidizing and acetogenic bacteria are presented in Table 2.

Table 2. Estimated free energy changes of selected biological reactions in anaerobic digesters (25 °C) and pH 7) — adapted from [48]

Reaction	$G°$ kJ
Acidogenic bacteria:	
$(C_6H_{10}O_5) + H_2O = C_6H_{12}O_6$	− 17.7
$C_6H_{12}O_6 = 3 CH_3CO_2 + 3 H^+$	−311.0
$C_6H_{12}O_6 + 2 H_2O = CH_3CH_2^- + H^+ + 3 CO_2 + 5 H_2$	−192.0
$C_6H_{12}O_6 = CH_3CH_2CH_2CO_2 + H^+ + 2 CO_2 + 2 H_2$	−264.0
$C_6H_{12}O_6 + 6 H_2O = 6 CO_2 + 12 H_2$	− 25.9
Acetogenic bacteria:	
$C_6H_{12}O_6 + 2 H_2O = 2 CH_3CO_2^- + 2 H^+ + 2 CO_2 + 4 H_2$	−216.0
$CH_3CH_2CO_2^- + H^+ + 2 H_2O = CH_3CO_2^- + H^+ + CO_2 + 3 H_2$	+ 71.7
$CH_3CH_2CH_2CO_2^- + H^+ + 2 H_2O = 2 CH_3CO_2^- + 2 H^+ + 2 H_2$	+ 48.3
$CH_3CH_2OH + H_2O = CH_3CO_2^+ + 2 H_2$	+ 9.7
$2 CO_2 + 4 H_2 = CH_3CO_2^- + H^+ + 2 H_2O$	− 94.9

The electrons generated during the microbial processes may be transferred to pyridine nucleotides and ultimately disposed of via proton reduction to form H_2. Certain incompletely oxidizing reactions proceed only at low H_2 concentrations accompanying formation of acetate as the major soluble product. Furthermore, the continuous removal of H_2 together with CO_2 then become important factors. When H_2 removal by methanogens is less efficient, H_2 blocks electron disposal by proton reduction and incompletely oxidizing bacteria must channel electrons to other disposal sites [49]. This results in increased production of reduced bioproducts such as propionate and butyrate, and is called the H_2 control point number [50]. The relationship between hydrogen concentration (partial pressures) and the end products of anaerobic catabolism is shown in Table 3.

When acidogenic bacteria are isolated from the digester and grown in axenic culture, there is no mechanism for interspecies H_2 removal. Such cultures may form bioproducts such as lactate and ethanol which are not usually produced in anaerobic digesters.

Some bacteria can catalyze a "back reaction" in which small amounts of acetate [64, 65, 66], propionate [67], butyrate [68] and succinate [69] are produced

Table 3. Relationship between hydrogen concentrations (partial pressure) and the end products of anaerobic catabolism (adapted from [51])

Culture Condition	Substrate	Hydrogen	Significance of Hydrogen Present	Ref.
Rumen: dialysis sac	Rumen fluid, H_2	10^{-6} M	Normal methanogenesis	[52]
Methenobacterium omelienskii	Ethanol media	0.5 atm	Ethanol degradation inhibited	[53]
Sewage digester sludge	Glucose pulses 4.4 g/L	0.01 atm	Propionic acid accumulation to 0.3 g/L, pH drop to 6.6, 3 days to acid conversion and pH recovery	[54]
Sewage digester sludge	Glucose pulses 8.8 g/L	0.02 atm	Propionate acid accumulation to 1.2 g/L, pH drop to 6.1, 5 days to pH recovery	[54]
Sewage digester sludge	Glucose pulses 13.3 g/L	0.50 atm	Butyric acid accumulation to 1.8 g/L, pH drop to 5.0, no recovery	[54]
Sewage digester	Acetic and propionic acid	not reported	Propionate to acetate degradation inhibited under hydrogen atmosphere	[55]
Clostridium cellobioparum pure culture	Glucose rumen fluid broth	0.3 atm	H_2 production reduced 50% accumulation butyrate, ethanol, lactate	[15]
Sewage digester sludge	Sludge, acetic acid	10^{-4} + 0.015 atm	Linear increase in propionic acid	[56]
Sewage digester sludge	Sludge, propionic acid, ethanol	0.005 atm / 0.07 atm	Propionate degradation inhibited, ethanol degradation inhibited	[57]
Sewage digester sludge	Wastewater sludge	0.18 atm	Propionic acid degradation inhibited	[58]
Thermoanaerobium brockii	Glucose	0.5 atm	Accumulation of lactate, ethanol H_2 production inhibited	[59]
Sewage digester sludge	Sludge	1.2 μM	Normal operation	[60]
Sewage digester sludge	Molasses and yeast	2×10^{-4} / 1.5×10^{-3}	Organic shock was caused H_2 increase followed by propionic acid accumulation	[61]
Cattle manure, batch digester	Cattle manure	addition of 70:30 H_2:CO_2	Propionic acid accumulation	[62]
Propionic acid enrichment	Propionic acid	6.5×10^{-5} / 4.3×10^{-5}	Normal operation at 8.2 day HRT / Normal operation at 14.5 days HRT	[63]

from H_2 and CO_2. Other studies indicate that mixed cultures produce volatile fatty acids from methanol [70, 71].

The methanogenic bacteria convert acetate and H_2/CO_2 into methane. These bacteria are very sensitive to oxygen and known as being the most strongly anaerobic.

Some of the notable species that have been classified are: *Methanobacterium formicicum, M. bryantic, M. thermoautotrophicum, Methanobrevibacter ruminantium, M. arboriphilus, M. Smithii; Methanococcus vannielli, M. voltae; Methanomicrobium mobile, Methanogenium cariaci, M. marinsnigri; Methanospirillum hungatei* and *Methanosarcina barkei* [72]. A more specific classification of methanobacteriaceae is presented elsewhere [73, 74].

All of these methanogens can use hydrogen as the sole electron donor for methanogenesis and growth. Carbon dioxide is reduced to methane when hydrogen is the substrate. Some species can use formate as a carbon and energy source by the conversion of the methyl carbon with its associated H-atoms into methane; the carboxyl group provides the source of CO_2 [48, 75]. When CO is used as the sole energy source, the thermophile *M. thermoautotrophicum* produces 3 moles of CO_2 and 1 mole of CH_4 from 4 moles of CO via the action of a CO dehydrogenase and hydrogenase with the fluorescent carrier F_{420} (cofactor F_{420}) as the electron acceptor [76]. The reduced F_{420} subsequently acts as the electron donor in the reaction reducing CO_2 to CH_4 [77, 78].

Some reactions of methanogenic bacteria are presented in Table 4.

Table 4. Reactions performed by methanogenic bacteria (adapted from [75] and [76])

Reaction	$G°$ kJ
$4 H_2 + HCO_3^- + H^+ = CH_4 + 3 H_2O$	-136
$CO_2 + 4 H_2 = CH_4 + 2 H_2O$	-131
$CH_3COO^- + H_2O = CH_4 + HCO_3^-$	$- 31$
$4 HCO_2^- + H_2O + H^+ = CH_4 + 3 HCO_3^-$	-134
$4 CH_3NH_3^+ + 3 H_2O = 3 CH_4 + 4 NH_4^+ + HCO_3^- + H^+$	-225

All types of anaerobic bacteria may live in symbiosis and all discussed reactions may be performed in one bioreactor. In a single-stage process, stable operating conditions prevail when a balance exists between the formation of acidification products (e.g. acetic acid, H_2/CO_2) and the subsequent formation of methane. With soluble carbohydrate-containing wastewaters such as those associated with food processing industries, the hydrolysis/acidification phase proceeds much more rapidly than the methanogenesis phase [81, 82]. Thus, in the single-stage treatment of these wastes, the tendency exists for the accumulation of hydrogen and acidic endproducts (e.g. acetic, propionic, butyric acids, ethanol) resulting in a drop in pH and the inhibition of the methanogenic phase [83]. When the process is performed in two stages the acidification stage can be operated independently of the methane stage, thereby enhancing process control and increasing the turnover rate of the whole treatment process [80, 81, 82, 84, 85, 86, 87].

3 Bioreactors

The anaerobic bed in reactors consists of all types of anaerobic bacteria which were described above although the proportion between them depends on the wastewater composition and on some process parameters, mainly pH and temperature [88].

For many years the disadvantage of anaerobic microbial processes was the low biomass production and consequently a very slow rate of digestion. The process had to be performed at very long retention times or even in batch culture [89]. Probably of major significance was the realisation that much of the performance of anaerobic digesters depends on the conversion of the acetic acid to methane and that the bacteria involved grew very slowly (mass doubling times of 8 to 80 days under many conditions). This agreed with the early empirical observation that hydraulic residence times of 60–100 days were required [46].

In conventional digesters, sludge is added in the zone where the sludge is being actively digested and the gas is being released. Fresh sludge is usually added to the system two or three times daily. As decomposition proceeds, three distinct zones develop. A scum layer is formed (Fig. 3) at the top of the digester, and beneath it are supernatant and sludge zones. The sludge zone has an actively decomposing upper layer and a relatively stabilized bottom layer. The stabilized sludge accumulates at the base of the digester. The scum layer is formed as the gas rises to the surface lifting sludge particles and other materials, such as grease, oils and fats. As a result of digestion, the sludge becomes more mineralized (the percentage of fixed solids increases), and the sludge layer becomes thicker because of gravity. In turn this leads to the formation of a

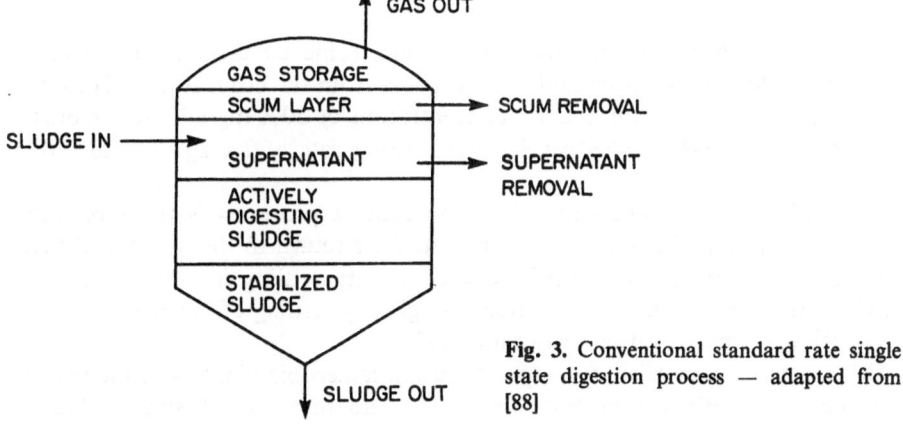

Fig. 3. Conventional standard rate single state digestion process — adapted from [88]

supernatant layer above the digesting sludge. As a result of the stratification and the lack of intimate mixing, not more than 50% of the volume of a standard rate single stage digester is used. This type of digester is still used in countries with a naturally warm climate. The different constructions of standard digesters are presented in Ref. [2].

The digestion could be accelerated by the increase of diffusion rate, biomass accumulation or both.

3.1 Anaerobic Contact Reactor

Mixing of the medium is the simplest method of increasing the diffusion rate. An anaerobic reactor with mixing, named High Rate Digester, is presented in Fig. 4. This type of digester does not have supernatant separation and a large amount of biomass is removed with the effluent. Usually, in this type of reactor, the reduction of wastes is fast [90].

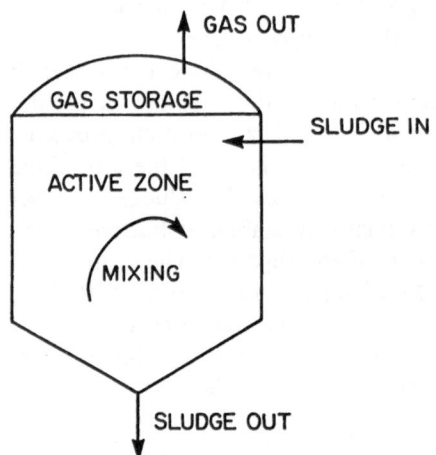

Fig. 4. High Rate Digester

   This method has been improved in the anaerobic contact process where a settler was added to separate and recycle solids from the effluent [91, 92] as it is shown in Fig. 5 [93]. With a COD concentration of 5000 mg L^{-1} at a biomass concentration of 5–10 g VSS per 1 L the loading rate of 2–6 kg COD $m^{-3} d^{-1}$ was possible [94].

   The major problem encountered in the contact processes is the separation and concentration of biomass flocs prior to their return to the digester. Several methods of separation were used, such as gravity settling in a thickener [95], gravity settling in a lamellar separator [96], gravity settling of polymer enhanced flocs [97], and centrifugation or flotation [92].

   Other devices have been developed to retain anaerobic biomass in the reactor and achieve an efficient biomass separation and recycling. A suitable barrier

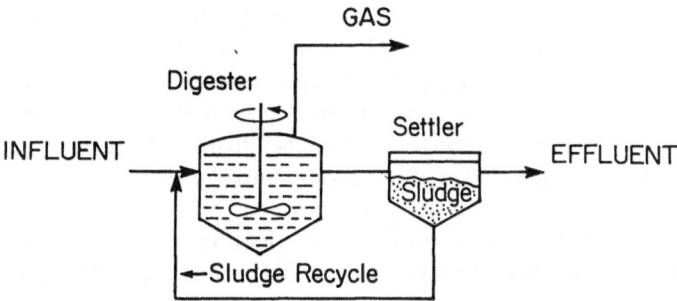

Fig. 5. Anaerobic contact process — adapted from [93]

maintaining anaerobic flocs within the reactor has been applied in anaerobic filters. Bacteria could be attached to each other. This is the case in the upflow anaerobic sludge blanket reactor (UASBR) where bacteria aggregate forming granules with excellent settling properties. The immobilization of bacteria on small inert support particles is realized in the fluidized/expanded bed reactor.

3.2 Anaerobic Filter

This reactor, which contains a solid support or packing material was developed by Young and McCarty [98]. The microorganisms in the anaerobic filter attach to the inert medium or become entrapped. In early applications of anaerobic filters the reactor was filled with stones used as packing (porosity about 0.4). Using low-medium strength wastes (COD concentration of 1000 to 6000 mg L^{-1}) the loading rates have not generally exceeded 6 kg COD m^{-3} d^{-1} for acceptable ($>70\%$) COD conversions. Biomass densities of 10–25 g VSSL^{-1} were obtained [99, 100, 101]. Later, highly packed materials with a porosity of 0.8–0.9 were used and high strength soluble wastes with a high loading rate of 10–20 kg COD m^{-3} d^{-1} were treated achieving 70% COD conversion and a good process stability [102, 103, 104].

The anaerobic filter tends to separate the microbial population. With upward flow, the acid forming bacteria act on the influent at the bottom of the digester, while the methane formers are found attached to the bed further along at location more distant from the influent. The bacteria remain stationary as the byproducts move through the bed [43, 105].

The main limitation of this process is the accumulation of solids in the packing material which may plug the reactor [106] The plugging may be caused by suspended solids from the waste water and by material precipitated from the waste (e.g. calcium carbonate). It is, however, mainly caused by methanogenic bacteria which have a natural tendency to attach themselves to surfaces and to each other.

The downflow anaerobic filter, usually named the Downflow Starionary Fixed Film Reactor, has a better plug resistance than the upflow anaerobic filter due to better mixing conditions at the top of the reactor. The influent stream entrains the flocs of methanogenic bacteria which can, in turn float to the top because of the attached gas bubbles or they can attach themselves to the wall of carriers in any other place within the reactor. In this way, a high concentration of biomass can be maintained in the reactor and higher loading rates may be used. The most common loading rate for the upflow anaerobic filters is in the range of 2–10 kg COD $m^{-3} d^{-1}$ but for downflow anaerobic filters it is in the range of 10–20 kg COD $m^{-3} d^{-1}$ [107]. However, the loading rate of 29.2 kg COD $m^{-3} d^{-1}$ for liquor from heat treated sewage digester sludge [108] and 39.2 kg COD $m^{-3} d^{-1}$ for piggery waste [109] have also been reported.

A schematic drawing of the anaerobic filter, with both upflow and downflow configuration, is presented in Fig. 6.

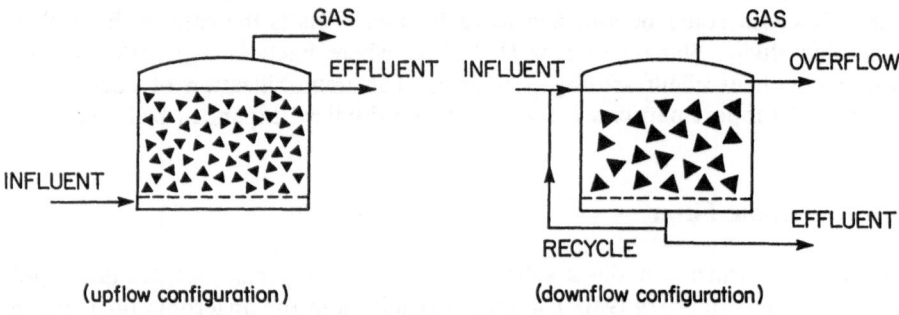

Fig. 6. Anaerobic filter reactors — adapted from [46]

3.3 Upflow Anaerobic Sludge Blanket Reactor (UASBR)

The concept of the upflow sludge blanket (USB) reactor was first investigated by Hemens et al. [110] and Cillie et al. [111] and has been developed by Lettinga and co-workers [112]. It can be described as a type of plug flow anaerobic activated sludge reactor.

In this type of anaerobic reactors the ability of bacteria to attach to each other is employed (refer to Sect. 4). Spherical aggregates composed of different types of bacteria can develop in the reactor when suitable process parameters are maintained. These aggregates settle very well. Their settling velocity depends on the size of aggregates [113] and usually is in the range of 20–90 m h^{-1} [114, 115] which can result in a low hydraulic retention time of even a few hours. Rising gas maintains granules in a more or less fluidised state and the resulting turbulence aids in detaching gas bubbles from granules in the upper part of the digester. The construction of reactors and the granulation methods for the UASB process have been proposed and developed by Lettinga and co-workers [70, 71, 112, 116, 117, 118, 119]. A schematic drawing of a reactor with a gas-liquid-solid separator is presented in Fig. 7. The UASB treatment process consists of the upward flow of the influent waste stream through a dense zone of anaerobic sludge, known as

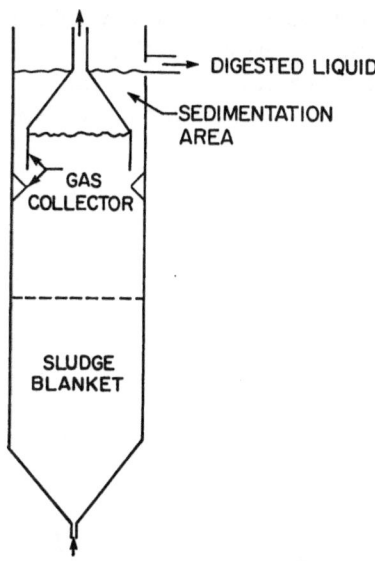

FEED

Fig. 7. Upflow anaerobic sludge blanket reactor — adapted from [107]

the sludge bed in which the organic components are metabolized and converted into gas. From the sludge bed, the waste stream enters a less dense zone of suspended flocs, known as a sludge blanket. The effluent leaves the reactor via a gas-liquid-solid separation device located at the top of the reactor.

3.4 Fluidized or Expanded Bed Reactor

In this type of reactor the bacteria attach to the small particles of the carrier, which may be unporous such as sand [120, 121], sepiolite [122], carbon [123] and different type of plastics [124, 125] or porous such as polyurethane foam [123, 126], pumice [127] and others.

The particles in the fluidized bed reactor can be very small (0.2–1 mm) and hence the surface area is very high (2000–5000 $m^2 m^{-3}$) [105] and then the biolayer thickness is very small and diffusion limitation is absent. By choosing the particle diameter and density, the settling rate of the biolayer covered particles can be up to 50 m h^{-1}. When a superficial liquid velocity of 10–30 m h^{-1} is maintained in the reactor a constant biomass concentration of 30–40 g VSS L^{-1} can be maintained [127].

A high superficial liquid rate, (promoted by recirculation), expands the bed. When the bed expansion is up to 30% the reactor is named the Expanded Bed Reactor. When expansion is between 30 and 100% the reactor is named the Fluidized Bed Reactor [93].

Very high loading rates of 20–30 kg COD $m^{-3} d^{-1}$ have been sustained with a variety of media and high strength soluble wastes [128].

Reactor performance depends very much on the evenness of the flow of the liquid and as a result the system of liquid distribution is very critical [129].

A schematic drawing of the fluidized/expanded bed reactor is presented in Fig. 8. Good informarion on the fluidized bed biofilm reactors for wastewater treatment is presented in [130].

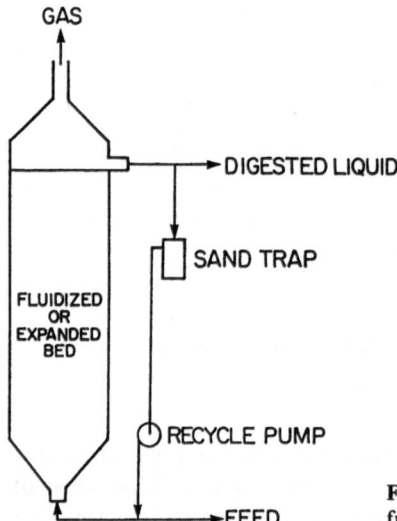

Fig. 8. Fluidized or expanded bed reactor — adapted from [93]

3.5 Discussion

The most important average values of process parameters for the types of reactors which were discussed above are presented in Tables 5 and 6. The actual values of these parameters depend strongly on the type of wastewater, pH and temperature.

Table 5. Reactor loading rates in the mesophilic range (adapted from [46]

Anaerobic reactor type	Typical loading rates kg COD $m^{-3} d^{-1}$
Continuous Stirred Tank Reactor (CSTR)	0.25–3.0
Contact Reactor (CR)	0.25–4.0
Upflow Anaerobic Sludge Blanket Reactor (UASBR)	10– 30
Anaerobic Filter (AF)	1– 40
Fluidised Bed Reactor (FBR)	1–100

Table 6. Hydraulic Retention Time (HTR) and Sludge Retention Time (SRT) at 35 °C for anaerobic digestion systems (adapted from [46])

Reactor Type	HRT Days	SRT Days	Organic Feedstock
CSTR	10–60	10–60	High SS; low to high strength
CR	12–15	20	Small but significant SS; low to moderate strength, but not carbohydrate rich
UASB	0.5–7	20	Low SS; low to rich strength
AF	0.5–12	20	Low SS; moderate strength
FBR	0.2–5	30	Low SS; moderate to very rich strength

A two stage process is recommended for wastewaters containing a large amount of soluble carbohydrates. As previously discussed, methanogenic bacteria grow more slowly than those in the nonmethanogenic group. Consequently, operating a first stage at short retention time promotes the selective washout of methanogens and a hydrolysis of the biomass feed into simpler components. Because methanogenic bacteria are absent with this digester, little methane production occurs. The incompletely oxidized product from the first stage is digested by the large methanogenic population in the second stage operating at the longer retention time and at significantly higher pH.

The two-phase concept can be employed with several reactor configurations provided the microbial retention time in the second stage exceeds that in the first stage digester. Because the pH in both of the stages must be precisely controlled, the well mixed reactors, such as the continuous stirred tank reactor or fluidised bed reactor, are preferred [131].

The UASB reactor seems to be the most promising construction for waste water treatment. The process is performed in one reactor, no support medium is required for attachment of the biomass which decreases the capital cost and minimizes the possibility of plugging. The energy requirement is also small because there is no mechanical mixing within the reactor, no recirculation of sludge and no high recirculation of effluent. Gentle recirculation of effluent is required to dilute too-strong wastes and to ensure proper hydraulic conditions within the reactor.

Other advantages of the UASB reactors are related to the properties of anaerobic granules which will be discussed in the next paragraphs.

4 Aggregation

4.1 The Phenomenon of Bacterial Aggregation

Many organisms have a tendency to attach themselves to surfaces or to each other [132, 133]. Freely suspended aggregates were observed for bacteria, yeasts, cellular slime molds, filamentous fungi and algae [134]. Generally, the microbial property for aggregation seems to be inducible. A natural tendency of *Methanosarcina mazei* toward aggregation in pure culture and a life cycle of this microorganism have been discovered [135, 136]. Small aggregates (up to 20 μm) were found to be characteristic for young cells. Mature culture formed coccal cysts of up to 100 μm and sometimes formed longer aggregates of up to several milimeters. Such a colony might disaggregate after physical stress and release individual cocci which can grow into the sarcinal form when placed in a fresh medium [133]. Micrographs illustrating the life cycle of *Methanosarcina mazei* are presented in [133]. Also, environmental conditions can cause a change in physiology so that the microbes change from a dispersed to an aggregated state. The aggregation may be advantageous due to interaction between adjacent organisms, such as commensolism, mutalism, parasitism and genetic exchange. Microbial growth within an aggregate or films may be selected by the hydraulics of the system as a possible protection mechanism. The nutrient uptake may also be increased comparing to freely dispersed cells [137]. Microbial starvation seems to be the main common condition which induces aggregation [137, 138]. Aggregates accumulated and were well maintained when the carbon source utilization from the medium was about 100% [139]. When inert particles are present, the aggregation depends significantly on the physical properties of surfaces [140, 141, 142, 143, 144].

A direct reason of the adhesion of cells to cells in microbial aggregates is supposed to be the production of an external layer of glycocalyx which was found to be selectively accumulated in UASB reactors [141, 145]. This substance is also believed to form irreversible bonds between cells and other surfaces [146]. Since the substance is generally found to be composed of polymers the name Extracellular Polymer Substances (EPS) is often used. The amount of EPS of 1–2% is reported for bacterial aggregates [147], however, in aggregates mainly composed of *Methanosarcina* sp. grown on propionate only 0.3% of EPS was found [148]. Polysaccharide [82] and polypeptides [113, 149] are supposed to promote bacterial adhesion and aggregation. The exact composition of EPS is still pourly defined, because the results and analysis depend on feed composition, process performance and methods of EPS extraction. 2 N NaOH [150], 1 N ammonium hydroxide [151], boiling [152], homogenization [153] are all recommended for EPS extraction but each of these methods provided quite different results [154].

Also, the nature of the microorganisms connection into aggregates is still not clear and a number of slightly different theories have been published.

Methanotrix and *Methanosarcina (Methanococcus)* are reported as the main species in anaerobic aggregates [107, 147, 155, 156, 157, 158, 159, 160, 161, 162, 163, 164].

It is known that *Methanosarcina* can form natural clusters but it does not attach itself easily to alien surfaces. The *Methanotrix* appear to be responsible for attachment to inert surfaces and they are supported to form the aggregate precursor under low acetate concentration [145]. When the loading rate is progressively and carefully stepped up the *Methanotrix* grow in rod like form and ensure the formation of the final dense compact aggregates which are attached to the surface of forming granules [145].

An alternative hypothesis [164] explains mainly the granule formation. According to this, the *Methanosarcina* clumps which are selected for the initial stages by the high acetate concentration are then colonized in central cavities by *Methanotrix*. This statement was supported by the observation that small, young granules have centres composed exclusively of *Methanotrix* with *Methanosarcina* on the outside. Subsequently, under the process conditions the more dense *Methanotrix* type granule is developed but the outer sarcinal layer is lost.

Both of these hypotheses predict that old compact aggregates should consist of *Methanotrix* which was confirmed by several observations [119, 123, 157, 160, 165, 166, 167]. A Transmission electron micrograph (TEM) of the bacteria incorporated in an anaerobic aggregate fed with synthetic feed composed of volatile fatty acids, is presented in Fig. 9.

Fig. 9. Transmission electron micrographs of bacteria incorporated in an anaerobic aggregate — adapted from [139]

More species of microorganisms have been discovered in floc-forming bacteria or in fresh biofilms [158, 168, 169, 170]. *Methanotrix* and *Methanosarcina* are also predominant but others such as *Desulfovibrio* [170], *Propionibacterium* [169], *Syntrophobacter* associated with *Methanobrevibacterium* and *Pelobacter* [169] were recognized by TEM.

The tendency of microorganisms to aggregate and adhere to the surfaces is reflected by some measurements of surface properties such as surface charge, hydrophobicity and the aggregation/flocculation behaviours. Changes in these parameters in response to various environmental conditions were observed and correlations sought between the cell surface interfacial properties (charge and hydrophobicity) and flocculation behaviour of anaerobic sewage sludge [171].

Eighmy et al. [168] have demonstrated that the process of bacterial adhesion is related to the negative surface charge density and to the relative hydrophobicity/hydrophilicity at the surface. It has been shown [172] that flocculation/ aggregation of activated sludge cells can be predicted using contact angle and zeta potential measurements.

The presence of divalent metal ions, which could act as a bridge between negatively charged groups on cell surfaces, have also proved to be important in the aggregation process. It was demonstrated that the concentration of calcium ions in the range of 40–100 mg L^{-1} in the waste stream enhances the rate of sludge aggregation [173]. The granules formed in the presence of large calcium concentrations settled 3–4 times faster than these at low calcium concentrations [174]. Anaerobic sewage cells were induced to aggregate with increasing concentration of Fe^{3+}, Al^{3+}, Ba^{2+}, Mg^{2+}, and Ca^{2+}. An increase in the tendency to aggregate was correlated with a reduction in the negative cell surface charge. The divalent cation which promoted the most rapid aggregation was barium while aluminium, a trivalent ion caused the most rapid settling overall [171].

Different cations may vary in their effectiveness in cell-cell binding. It was observed that there are striking differences in the effectiveness of various cations in promoting cellular adhesion or aggregation. These differences are primarily due to the selectivity of the ion exchange reaction of polymers (such as the bacterial surface), which result from variations in the charge to radius ratios of the cations. It is apparent therefore, that the strengths of various types of bonds produced by different cations will affect the resultant aggregation ability of the cells. The higher atomic weight ions also have a more pronounced effect on the floc density than do lower atomic weight ions, thus increasing the sedimentation rate [175].

It was found that wastes having an ammonium nitrogen concentration of approximately 1000 m L^{-1} suppress the aggregation or granulation of the sludge and enhance the development of a bulky sludge. It was discounted that the high concentration of ammonium nitrogen simply inhibits the methanogenic bacteria, but univalent cations in the form of the ammonium ion may have increased the electrical charge of the biological colloids (i.e. flocs) associated with the granulation phenomenon [119].

4.2 Granulation in UASB Reactors

During the early phase of the start-up, growth of the desired acidogenic, acetogenic and methanogenic organisms take places both as an attached and a dispersed biomass.

Through a slow and gradual selection process the dispersed growing organisms are washed out from the reactor, while biomass aggregates, which consist of microorganisms either attached to each other or attached to support particles are retained in the reactor. The selection is based on the minor differences in the settling properties (density) of free organisms and bacterial agglomerates. The selection pressure originated from both the hydraulic loading rate (or dilution rate) and the gas loading rate. The reactor height to diameter ratio determines both the superficial liquid and gas upflow velocities in the reactor.

In the first week of a start-up of the laboratory and pilot plant UASBR the upflow velocity is generally maintained at a very low level (0.1 to 1.0 m d^{-1}) although in full scale reactors it is frequently considerably higher (30–50 m d^{-1}) [176].

The upward moving fluid and gas bubbles produced by bacteria cause, directly after the start-up, an expansion of the sludge bed [177]. Thus, poorly settleable light materials such as colloids and dispersed growing biomass are washed out from the reactor from the very beginning. The heavy sludge ingredients tend to concentrate increasingly in the lower part of the reactor, while in the upper part more voluminous sludge is accumulated. Heavy bacterial consortia, which are collected in the lower part of the reactor, develop into aggregates while dispersed and voluminous bacteria are washed out of the system.

Granulation may occur at the range of temperature which is suitable for methanogenic bacteria. The granules developed from a very poor sludge at a temperature of 30 °C [116] and the thermophilic range temperatures (55 °C) were appropriate for granules formation [158]. The effluent quality after digestion at 55 °C was comparable to that at 30 °C at similar loading rates [178] although for ungranulated anaerobic sludge the loading rates applicable at 54 °C were 2.4 times as high as those at 38 °C [179].

The pH values between 6.6 and 7.6, which are characteristic for methanogenic bacteria, are recommended for developing and cultivation of anaerobic granules [83, 180]. Lower values such as 5.7 [158] and 6 [181] for thermophilic bacteria have also been reported.

When the temperature and pH are fixed at the proper range, the granulation in an UASB rector depends on process parameters such as the loading rate (kg COD m^{-3} d^{-1}), specific loading rate (kg COD (kg VSS)$^{-1}$ d^{-1}), upflow velocity (m h^{-1}) and the amount and the type of the seed sludge i.e. with respect to its specific activity, its settleability and the nature of the inert fraction.

The granulation of anaerobic sludge in organic rich wastes can occurr when the following restrictions are observed [119]:
— amount of seed sludge is 10–20 kg VSS m^{-3},
— initial specific loading rate is 0.05–0.1 kg COD (kg VSS)$^{-1}$ d^{-1},
— the substrate loading rate should not be increased unless all volatile acids are well degraded (i.e. greater than 80 % removed for medium strength wastes),
— voluminous sludge should be allowed to wash-out and the heavy part of the seed should be retained,
— long period of overloading or underloading with respect to the specific substrate loading rate should be avoided.

The granulation process is not restricted to methanogenesis. A granule formation (granule diameter up to 5 mm) in an upflow sludge bed reactor which served for denitrifying nitrified final effluent mixed with settled domestic sewage has been reported [182, 183]. These granules were formed only at a high upflow velocity (8 m h^{-1}) and contained up to 50% CaCO$_3$, but were rather unstable.

The granulation was also observed in the acidification of glucose in an USB reactor at residence time shorter than 1 hour at a glucose concentration of 10 kg m^{-3} and shorter than 16 hours at 50 kg m^{-3}. The granules formed were about 1 mm in diameter and were unstable under unfed conditions.

The fact that methanogenic granulation takes place at higher residence times and that more stable granules are formed with anaerobic waste water treatment (granular methanogenic sludge remains well conserved under unfed conditions for several years) gives evidence that the mechanism of the formation of methanogenic granules differs from that of denitrifying and acidifying bacteria.

Through the increasing practical experience with UASB reactors it became obvious that the composition of the wastewater plays a rather dominant role. Wastewaters as indicated in Table 7 gave good results, while problems arose with rendering waste, distillery waste and corn starch waste. However, with all these types of waste, granulation of the sludge ultimately occurred.

Table 7. Process data of some UASB reactors initially seeded with digested sewage sludge

Wastewater	Methanogenic activity kg CH$_4$(kg VSS)$^{-1}$ d^{-1}	Loading rate kg COD m^{-3} d^{-1}	HRT h	Ref.
VFA mixture	2.2	60	1	[119]
Yeast Factory	0.7–0.9	7–14	17	[119]
Sugar Beet Factory	1.3	14	7	[184]
Potato Processing Factory	1.0–1.4	30–40	4	[119]

4.3 The Influence of Process Parameters on Granules Maintenance

The granules are maintained in a dense slightly fluidized bed which essentially governs the process performance and the activity of the anaerobic digestion system. It has been found that the process parameters and the environment in which the granules are suspended have a profound influence on the reactor performance.

A temperature of between 30 and 55 °C which was found to be beneficial for formation of granules, is also optimal for their maintenance.

The organic and volumetric loadings exert also a profound effect on the granular system and its performace. It is therefore important to establish optimum and maximum loading rates at which UASB reactors can maintain a sustained high efficiency.

Once an active granular biomass is formed the operational loading rate of up to 1.5 kg COD (kg VSS)$^{-1}$ d^{-1} or even up to 1.8 kg COD (kg VSS)$^{-1}$ d^{-1} can be maintained (113). Trudell et al. [185] overloaded the system for a short period of time to a level of 2.4 kg COD $(kg\ VSS)^{-1}\ d^{-1}$ and granules tolerated this shockload. It was found that granules can adapt to gradually increasing the loading rate and then the specific loading rate of 1.5 kg COD $(kg\ VSS)^{-1}\ d^{-1}$ can be maintained for a very long period of time with COD conversion of over 90% [186]. In batch culture the initial COD concentration up to 8000 mg L^{-1} did not influence the activity of the granules. Higher initial COD concentrations inhibited the activity and a disintegration of granules was observed although the granules could digest the wastes with COD concentration of up to 24,000 mg L^{-1} [187].

At low loading rates, in continuous culture, a hollow core may develop within the granules which makes them less dense than before, which causes their flotation and easy wash-out of the reactor [186]. Such a phenomenon may develop in a large part of the bed which may drastically reduce the bed effectivity. An electron micrograph of a granule with a hollow core is presented in Fig. 10.

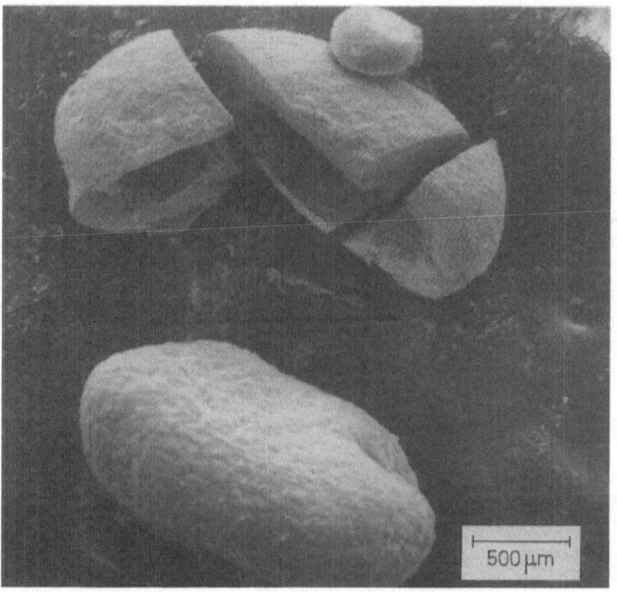

Fig. 10. Scanning electron micrograph of a granule with a hollow core — adapted from [166]

There is very little information concerning differences in the colour of the granules. It is possible that the colour of granules may give an early signal about a change in their activity. It was reported [169, 188] that the granules which formed a well defined bed with a clear supernatant and without fluffy conclomerates were black although some whitish conglomerates were always present. No significant microscopic difference was found in their bacterial composition [188]. Bochem et al. [161] reported that in partially cross-sectioned granules (1–3 mm in diameter) one can recognize cells consisting of two regions: a densely packed outer area and a loosely packed central area resembling a lysis zone. This latter area was comprised of ovoid cells (1.0–2.0 μm in diameter) with intercellular spaces of approximately 0.1 μm. These intercellular spaces were either completely transparent or composed of weakly stainable matter. During the examination of the colonization of sand grains by bacteria the numerical cell densities within the aggregates ranged from 10^{12} in 1 mL at the periphery to very low values in the centre [162]. Microscopic evidence for "hollowing" of large aggregates by autolysis has been presented, and it was suggested that granules lysed within large aggregates due to substrate insufficiency [189]. Conceivably, such "hollow" structures are more susceptible to shear stresses which may occur within the reactor. The disappearance of large aggregates may also have resulted from their disintegration.

Granules exposed to high calcium concentrations during a long period of time changed their colours significantly from almost black to grey or even white [190].

Table 8. Percent fraction of elements in cross-sectioned black, grey and white anaerobic granules — EDAX analysis adapted from [190]

	Element	Surface	Periphery	$^1/_2$ Radius	Core
A. Black					
	Ca	19.76	13.26	19.44	1.83
	Fe	7.95	5.59	7.40	5.75
	Cu	1.75	3.16	2.67	2.81
	S	14.13	19.06	18.53	15.47
	P	19.66	21.45	15.47	15.62
	Si	14.22	13.63	14.08	16.48
	Al	19.62	18.42	19.94	20.61
	K	0.58	1.18	0.58	0.73
	Na	1.57	3.70	1.07	0.48
	Cl	0.76	0.55	0.70	0.22
B. Grey					
	Ca	18.72	15.31	17.49	18.24
	Fe	6.22	8.04	7.62	5.69
	Cu	4.31	9.06	3.58	3.78
	S	34.87	28.62	18.87	17.47
	P	27.22	19.51	15.90	16.60
	Si	2.39	3.84	13.91	15.35
	Al	1.55	3.80	18.58	19.05
	K	3.56	4.90	2.16	2.56
	Na	0.11	4.70	1.30	0.66
	Cl	1.01	2.22	0.60	0.32
C. White					
	Ca	12.18	18.39	14.98	n/a
	Fe	3.86	4.60	4.49	n/a
	Cu	6.17	7.97	8.64	n/a
	S	31.17	26.18	35.17	n/a
	P	23.87	23.93	23.39	n/a
	Si	1.89	3.81	1.43	n/a
	Al	1.05	1.07	0.00	n/a
	K	10.94	6.53	6.36	n/a
	Na	7.30	5.12	2.85	n/a
	Cl	1.58	2.42	2.67	n/a

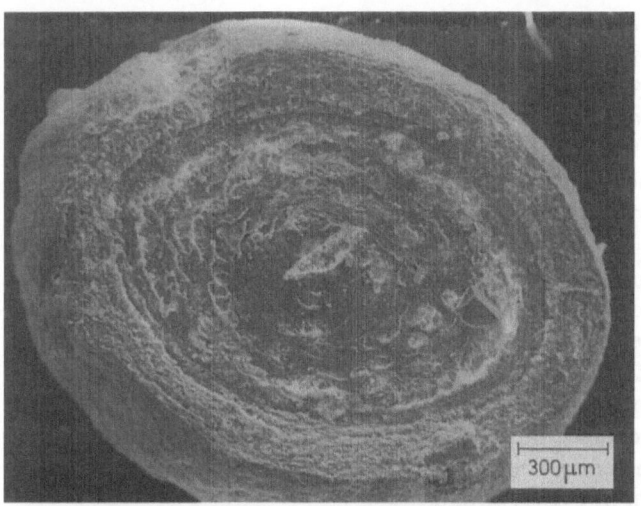

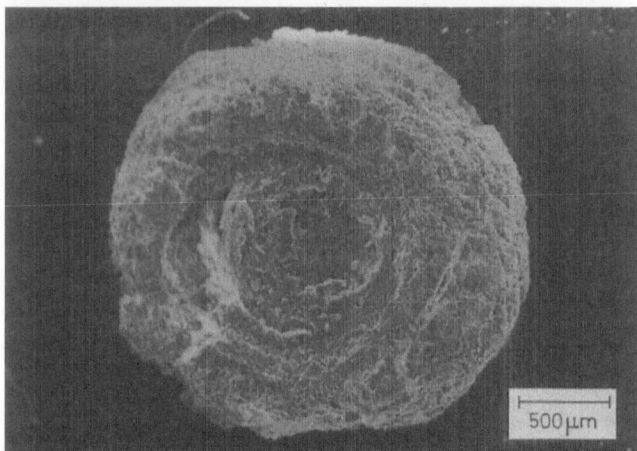

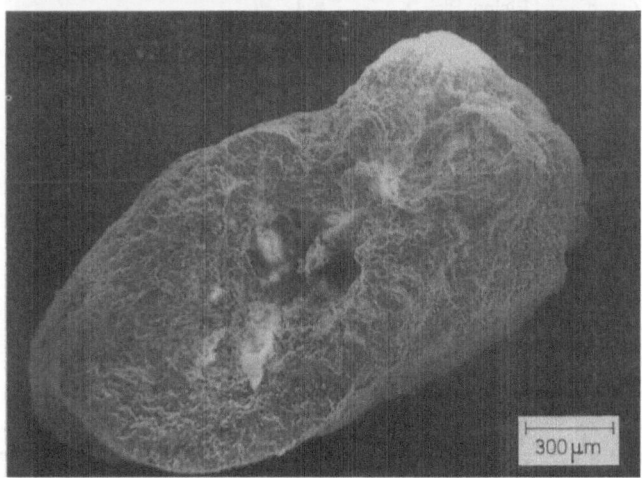

Fig. 11. Scanning electron micrograph of black, grey and white granules from an UASB reactor — adapted from [190]

Generally, the black granules were more firm than the grey ones, which were more firm than the white ones. The black granules had smooth surfaces, and were spherical or slightly ovoidal. The bacterial cells present were either filamentous or short curved rods. The surface of the grey granules was more irregular, having a more coarse texture. White granules were the most irregular, their surfaces had fissures and these granules were often broken. The X-ray analysis performed on four regions of the granules refered to as surface, periphery, half radius and core zones, showed that the black granules contained a large concentration of silicon and aluminium throughout. The grey granules contained similar concentrations of these elements in their core, but much less toward their surface. White granules contained uniformly low levels of these two elements. Calcium concentrations were generally independent of the granule colour perhaps due to the high concentration of calcium (100 mg Ca^{++} in 1 L) in the waste water to be treated. The elemental compositions in four granule zones for black, grey and white granules are presented in Table 8, a, b, c and the scanning electron micrographs (SEM) are shown in Fig. 11 a, b, c.

Another reason for occasional loss of granules from the reactor is their possible disintegration by mechanical shear within the reactor and following wash-out. Well formed granules from the UASB reactors usually settle very quickly [119]. Lettinga et al. [191] reported granule settling velocities in the range

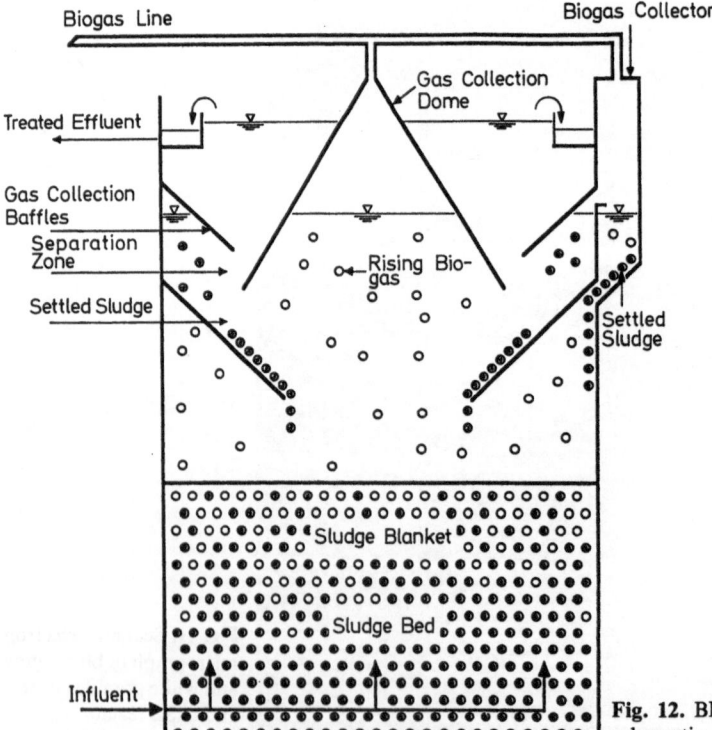

Fig. 12. BIOPAQ UASB reactor schematic — adapted from [202]

of 2–90 m h^{-1}. Usually the settling velocity is related to the size of the granules [169] and increases proportionally to their size. Li and Garnaczarczyk [192] reported that there is a linear or fractional power relationship between the individual particle's floc-settling velocity and the floc size. However, this settling velocity may be independent of the diameter of the floc for d > 2 mm [189].

When granules are exposed to mechanical stress the disintegration is observed. Tramper et al. [193] investigated the behaviour of granulated sludge in a stirrer tank and a Couette vessel at various stirrer and rotational speeds. A strong increase in abrasion was found when mechanical stress increased above a certain, fairly low level. The influence of pump pulsation on the shape of granules was also reported [194].

The strong influence of upflow velocity on granule size and settling characteristics was reported by Kosaric et al. [115]. The granules were exposed to the upflow velocities from 0.25 to 1.5 m h^{-1}. At the highest velocity no accumulation of granules was observed and granules were uniform in size. At the lowest velocity, a strong accumulation of granules was observed. Recent investigations show that not only the upflow velocity in the reactor but also the velocity in the inlet ports and the configuration of these ports play a very important role [195]. On the other hands it was also reported that larger granules occured at a higher liquid upflow velocity (2 m h^{-1}) in UASB reactors while smaller granules appeared at lower velocity (1 m h^{-1}) [196].

4.4 Industrial Application of Microbial Aggregates for Wastewater Treatment in UASB Reactors

The concept of the UASB reactor in which bacteria can aggregate into firm, mechanically resistant and dense granules is very attractive for industrial applications. Additionally, such a reactor does not require expenses for reactor

Table 9. Early research in pilot scale UASB reactors using granular sludge

Type of wastewater	Maximum loading rate kg COD m^{-3} d^{-1}	COD reduction %	Ref.
Sugar beet factory	30–32	75	[114]
Potato processing	40	84	[198]
Potato starch prod.	30	75	[199]
Vegetable canning	10–20	60–80	[198]
Yeast factory	14	70	[198]
Dairy factory	15	80	[198]
Slaughterhause	10	55	[198]
Rendering plant	60	63	[198]
Distillery waste	11–17	45–65	[85]

packing or biomass support media and the energy consumption is relatively low
[197]. Succesful investigation at a laboratory and at a pilot scale were performed
almost simultaneously in the early 1980s. Several research groups from the
Netherlands conducted thorough experiments on granulation (methanogensis
and acidification), start-up conditions and liquid flow in the reactor. These investigations resulted in fully successful pilot plant processes as shown in Table 9.
Industrial installation started soon after and in the early 1980s a number of full.
scale reactors were set in motion. Sugar beet wastes, potato starch wastes and
dairy wastes were mainly digested in these reactors (Table 10). Further on,
wastewaters containing soluble organics were treated in UASB reactors.

Table 10. Early full scale UASB reactors

Type of waste	Waste strength kg COD m^{-3}	Loading rate kg COD m^{-3} d^{-1}	Reactor volume m^{-3}	Conv. %	Ref.
Sugar beet	1–2.6	10	800	88–93	[85]
	1.6	14–32	200	87–95	[114]
	4.2	14	200	90	[184]
Starch	2.2	11	1800	85	[200]
Dairy waste	2.5	6–13	400	60–67	[201]

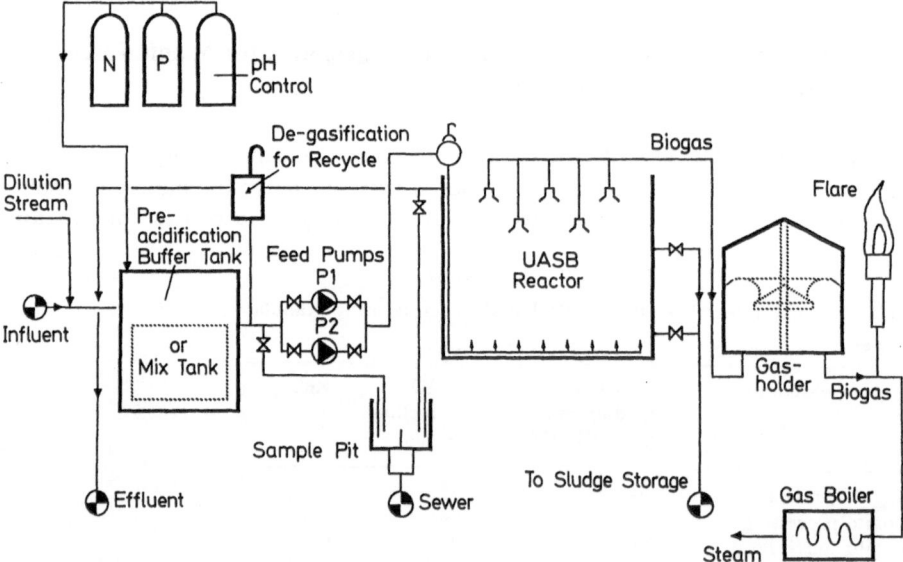

Fig. 13. Anaerobic treatment system adapted from [202]

An industrial scale UASB reactor (produced by Paques BV in the Netherlands
and Paques-Lavalin in Canada) is presented in Fig. 12. The construction is very
simple and all flows except the feed stream are by gravity. No circulation pumps,

centrifuges and packing materials (except for a gas-liquid-solid separator) are employed. The reactor consists of a corrosion resistent tank which incorporates a unique 3-phase settler to separate the sludge, biogas and effluent. This settler is located at the top of the reactor and is designed for specific COD reactor loadings and a hydrodynamic throughput. A flow distribution network is located at the base of the reactor. This network is designed to distribute the flow evenly throughout the bottom of the reactor. This eliminates short-circuiting and promotes the proper formation of the granular sludge which is a critical factor in reactor operation. This distribution network also facilitates cleaning thereby eliminating potential plugging problems. The biogas produced by the bacteria is in a form of small bubbles which float upward through the sludge bed/blanket zones and provide a natural mixing action. When the biogas reaches the top of the reactor, it is removed by gas collectors. A gas free zone above the collector makes possible the settling of finely dispersed solids to the reactor bottom. The UASB reactor is part of the anaerobic treatment system shown in Fig. 13. In this system the preacidification/buffer tank is located upstream from the UASB reactor. This

Table 11. Rough cost estimate of anaerobic treatment in 1000 and 5000 m³ UASB reactors (calculated in [202])

1.	Assumption made in the estimate.	
	COD load	10 and 15 kg m^{-3} d^{-1}
	treatment efficiency	90% COD reduction
	methane yield	0.9 kg COD-CH$_4$ per kg COD removed
	interest and redemption	15% of the capital costs
	maintenace and renewals	2% of the capital costs
	energy requirements	10% of the methane production
	investment costs 1000 m³	$ 500,000–700,000
	investment costs 5000 m³	$ 2,000,000–3,000,000

		1000 m³ plant	5000 m³ plant
2.	Operation costs (in $ 1000•)		
2.1.	Continuous operation (365 day a year, 24 hrs a day)		
	interest + redemption	75–112.5	300–450
	maintenance + renevals	10–15	40–60
	labour + supervision	15	40
	analyses + control	15	40
	total costs	115–147.5	420–590
	Costs of methane gas ($ per m³ STP)		
	1° load 15 kg COD m ×³ d^{-1}	0.08–0.105	0.06–0.085
	2° load 10 kg COD m^{-3} d^{-1}	0.125–0.160	0.09–0.125
2.2.	Seasonal operation (3 month a year, 24 hours a day)		
	interest + redemption	75–112.5	300–450
	maintenance + renewals	3–5	15–20
	labour + supervision	5	15
	analyses + control	5	15
	total costs	88–127.5	335–500
	Costs of methane gas ($ per m³ STP)		
	1° load 15 kg COD m^{-3} d^{-1}	0.25–0.36	0.19–0.29
	2° load 10 kg COD m^{-3} d^{-1}	0.38–0.55	0.29–0.43

Table 12. BIOPAQ industrial reactors installed by Paques-Lavalin, Toronto, Canada (adapted from [202])

Company	Industry	Reactor Volume m³	Flow m³ d⁻¹	Influent COD g L⁻¹	Volume Loading µg COD m⁻³ d⁻¹	Temperature °C	Efficiency of COD Removal %	Gas Production m³ d⁻¹	CH₄ v v⁻¹	Year Installed
Kuibo	french fries	300	380	5.0	6.4	33–37	70	500	80	1981
Fri d'Or	french fries	1300	920	9.0	6.4	33–37	85	3000	75	1983
Wheat Starch	starch	2200	840	20.0	7.5	33–37	85	3800	80	1983
Roermond Papier	paper	1000	2900	3.5	10.0	30–40	75	3600	80	1983
Residentie Slachthuis	slaughterhouse	650	1040	4.0	6.4	33	75	1400	75	1983
Ruiten Troes	cannery	375	1125	4.2	8.0	32	85	1000	75–80	1984
Vewco	licorice	50	36	18.0	13.1	32	90	210	75	1984
Bavaria	brewery	1400	6000	1.6	6.8	20–24	80	1500	85	1984
Celtona	paper	700	2900	1.2	5.0	20–25	60	800	80	1984
Usina, Sao Luiz, S.A.	distillery	120	40	45.0	15.0	35	80	864	70	1984
Industrie-Water Eerbeck	paper	2200	9600	1.3	5.7	25	68	2250	84	1985
Davidson	paper	1600	6000	3.0	11.2	35	65	3900	75	1986
Usina Sao Joao leboa vista	distillery	1500	750	30.0	15.0	35	80	10800	70	1984

tank is required to buffer peak hydraulic loads, peak organic loads, extremes in temperature fluctuation, and to acidify the wastewater (typically a minimum of 30 percent acidification is required). This tank can also accept the reactor storage system. The excess granular sludge from the reactor is usually used for reseeding, if required, or as a seed sludge to start up other UASB reactor installations. Shown in Table 11 is a review of process parameters maintained in effluent for buffering purposes and chemical nutrients to facilitate preacidification, if required. Biogas from the UASB reactor is discharged from the top of the reactor and is collected in a common header for discharge to the biogas cleaning and recently installed BIOPAQ Watewater Treatment System (Paques BV in Europe and Paques Lavalin in Canada) [204].

These and other anaerobic treatment systems have been operated succesfully for many years [93]. More than half of the full scale UASB reactors have been installed in the Netherlands. The rough cost of such installations, with capacities of 1000 and 5000 m³, calculated by Lettinga and Hulshoff Pol [203] is presented in Table 12.

One of the main problems in the operation of industrial reactors is in "degranulation" e.g. disintegration of granules during operation. The disintegrated granules/flocs lose their high settling ability and are consequently washed out of the reactor making the system totally inoperable. Replacement and/or redevelopment of granules is a costly and timely undertaking. The equilibrium between the loading rate, biomass concentration and the hydraulic conditions in the reactor seem to be responsible for the stability of the granules in UASB reactors [115, 186]. Also high concentration of wastes within the reactor (more than 8,000 mg COD L⁻¹) promote granule disintegration [205]. Also toxic and/or hazardous compounds may destroy a high efficiency methanogenic bed.

For the above reasons a two-step treatment process, containing two reactors was proposed. The first reactor is for liquefaction and/or acidification and the second reactor for methanogenesis. For example, for the potato starch treatment plant in "De Krim" [203] a 1700 m³ acidogenic reactor has been placed in front of the 5500 m³ methanogenic reactor. The main objectives of the first reactor were to remove SO_3^{2-} and proteins. The two phase process is also recommended even in the case of merely soluble types of wastes to provide a higher stability of the process [206, 207, 208] but the benefits are questionable and investment costs for two-step processes may exceed significantly that of a one step process.

5 References

1. Verstraete W (1983) Biomethanation of wastes: perspectives and potentials, In: Biotech 83: Proc Int Conf on the Commercial Applications and Implications of Biotechnology. Online, London, p 725
2. Stafford DA, Hawkes DL, Horton R (1980) Methane production from waste organic matter. CRC Press Boca Raton, FL
3. Eastman JA, Fergusson JK (1981) Journal WPCF, 53: 352
4. Hobson PN, Wallace RJ (1982) CRC Crit Rev Microbiol 9: 162
5. Hobson PN, Wallace RJ (1982) CRC Crit Rev Microbiol 9: 253

6. Hungate RE (1950) Bacteriol. Rev. 4: 1
7. Hungate RE (1957) Can. J. Microbiol. 3: 289
8. Johnston EA, Sakajoch M, Madia A, Demain AL (1981) Appl. Environ. Microbiol. 43: 1125
9. Ng TK, Weimer PJ, Zeikus JG (1977) Arch. Microbiol. 114: 1
10. Ng TK, Zeikus JG (1981) Appl. Environ. Microbiol. 42: 231
11. Groleau D, Forsbrg CW (1981) Can. J. Microbiol. 27: 517
12. Latham MJ, Booker BE, Pettipfer PL, Harris PJ (1978) Appl. Environ. Microbiol. 35: 156
13. Hungate RE (1945) J. Bacteriol. 53: 631
14. Saddler JN, Khan AW (1980) Can. J. Microbiol. 26: 760
15. Chung KT (1976) Appl. Environ. Microbiol. 31: 342
16. Dehority BA, Scott HW (1967) J. Dairy Sci. 50: 1136
17. Dehority BA (1966) J. Bacteriol. 91: 1724
18. Bryant MP, Small N (1956) J. Bacteriol. 72: 16
19. Salyers AA, West SEH, Vercellotti JR, Wilkins TD (1977) Appl. Environ. Microbiol. 33: 529
20. Salyers AA, Ghardini F, O'Brien M (1981) Appl. Environ. Microbiol. 41: 1065
21. Healy JB Jr, Young LY (1977) Appl. Environ. Microbiol. 35: 216
22. Healy JB Jr, Young LY (1979) Appl. Environ. Microbiol. 38: 84
23. Healy JB Jr, Young LY, Reinhard M (1982) Appl. Environ. Microbiol. 39: 436
24. Schink B, Pfenning N (1982) Arch Microbiol. 133: 195
25. Rode LM, Sharac-Genther BR, Bryant MP (1981) Appl. Environ. Microbiol. 42: 20
26. Bryant MP, Small N (1956) J. Bacteriol. 72: 16
27. Dehority BA (1969) J. Bacteriol. 99: 189
28. Ward K, Seib PA (1970) Cellulose lichenin and chitin. In: Pigman W, Horton D (eds). The carbohydrates, chemistry and biochemistry, Academic, New York, p. 413
29. Schink B, Ward JC and Zeikus JG (1981) Appl. Environ, Microbiol. 42: 526
30. Sheiman MJ, MacMillan JD, Miller R and Chase T (1976) Eur. J. Biochem. 64: 565
31. Hungate RE (1957) Can. J. Microbiol. 3: 289
32. Hamlin LJ, Hungate RE (1956) J. Bacteriol. 72: 548
33. Hungate RE (1960) Bacteriol. Rev. 24: 353
34. Nakumura LK, Crowell CD (1979) Dev. Ind. Microbiol. 20: 531
35. Hobson PN, Mann SO (1961) J. Gen. Microbiol. 25: 227
36. Hazelwood GP, Dawson RMC (1975) J. Gen. Microbiol. 89: 163
37. Hazelwood GP, Dawson RMC (1979) J. Gen. Microbiol. 112: 15
38. Blackburg TH (1968) J. Gen Microbiol. 53: 27
39. Hazelwood GP, Jones GA, Mangan JL (1981) J. Gen. Microbiol. 123: 223
40. Hazelwood GP, Edwards R (1981) J. Gen. Microbiol. 125: 11
41. Sleat R, Mah R (1987) Hydrolytic bacteria In: Chynoweth DP, Isaacson R (eds), Anaerobic Elsevier, London
42. Mosey FE (1982) Mathematical modelling of the anaerobic digestion process: regulatory mechanisms for the formation of short-chain volatile acids from glucose. IAWPR-Seminar on anaerobic treatment, June, Copenhagen. Denmark
43. Hense M, Harremoest P (1983) Wat. Sci. Tech. 15: 1
44. Trosch W, Keller E, Stephan E, Chmiel H (1983) Microbial generation of Methane from a proteinous industrial waste water comparison between single and two-stage processing. Proc. of the Europ. Symp. Anaerobic Waste Water Treatment, 23–25 Nov. Noord-wijkerhout, Netherlands, p 117
45. Sorensen J, Christiensen D, Jorgensen BB (1981) Appl. Environ. Microbiol. 42: 5
46. Stronach SM, Rudd T, Lester JN (1986) Anaerobic digestion processes in industrial wastewater treatment, Biotechnology Monographs, vol 2, (Aiba S, Fan LT, Fiechter A and Schugerl K (eds)), Springer, Berlin Heidelberg New York
47. Doelle HW (1981) Basic metabolic processes. In: Rehm HJ and Reed G (eds). Biotechnology: vol 1: Microbial Fundamentals. Verlag Chemie, Weinheim
48. Klass DL (1984) Methane from anaerobic fermentation, Science, 223 (4640): 1021

49. Wolin MJ (1982) Hydrogen transfer in microbial communities. In: Bull AT and Slater JH (eds), Microbial interaction and communities. New York and London, Academic Pres
50. Iannoti EL, Kafkewitz P, Wolin MJ, Bryant MP (1973) J. Bacteriol. 114: 1231
51. Harper SR, Pohland FG (1985) Biotechnol. Bioengn. 28: 585
52. Hungate RE (1967) Arch. Microbiol. 59: 20
53. Bryant MP, Wolin EA, Wolin MJ, Wolfe RS (1967) Arch. Microbiol. 59: 20
54. Sykes RM (1970) Hydrogen production in the anaerobic digestion of sewage sludge, Ph. D. Thesis, Purdue Univ. Lafayette IN
55. Smith PH, Shuba PJ (1973) Terminal anaerobic dissimilation of organic molecules. In: Procedings of the Bioenergy Research Conference, Amherst, MA, June
56. Kaspar HF, Wuhrmann K (1978a) Appl. Environ. Microbiol. 36: 1
57. Kaspar HF, Wuhrmann K (1978b) Microbiol. Ecol. 4: 241
58. Smith MR, Mach RA (1980) Appl. Environ. Microbiol. 39: 993
59. Ben Bassat A (1980) Metabolic control for microbial fuel production during termophilic mannual of determitive bacteriology, part 13, Wiliams and Wilkins, Baltimore, p 472 fermentation of biomass. In: Proceedings of 4th Symposium on Energy from Biomass and Wastes, Lake Buena Vista FL
60. Robinson AJ, Tiedje KM (1982) Appl. Environ. Microbiol. 44: 1374
61. Barnes D et al. (1983) Influence of organic shock loads on the performance of an anaerobic fluidized bed system. Proceedings of the 37th Purdue Industrial Waste Conference. Purdue Univ
62. Boone DR (1982). Appl. Environ. Microbiol. 43: 57
63. Heyes RH, Hall RJ (1983) Appl. Environ. Microbiol. 46: 710
64. Fontaine FE, Peterson WH, McCoy E, Johnston MJ, Ritter GJ (1942) J. Bacteriol. 43: 701
65. Kerby R, Zeikus JG (1983) Curr. Microbiol. 8: 27
66. Balch WE, Schobertth S, Tanner RS, Wolfe RS (1977) Int. J. Syst. Bacteriol. 27: 355
67. Braun B, Mayer F, Gottschalk G (1981) Arch. Microbiol. 128: 288
68. Mackie RI, Bryant MP (1981) Apll. Environ, Microbiol. 41: 1363
69. Laanbroek HJ, Abee T, Voogd IL (1982) Arch. Microbiol. 133: 178
70. Lettinga G, van der Geest AT, Hobma S, van der Laan (1979) Water Res. 13: 725
71. Lettinga G, de Zeeuw W, Ouborg E (1981) Water Res. 13: 725
72. Balch WE et al. (1979) Microbiol. Rev. 43: 260
73. Bryant MP (1974) Methane producing bacteria. In: Buchann RE, Gibbons NE, Cowan ST Holt JG, Liston J, Murray RGE, Niven CF Ravin AW, Stanier RY (eds) Bergy's
74. Ferguson T, Mah R (1987) Methanogenic bacteria. In: Chynoweth DP and Isaacson R (eds) Anaerobic digestion of biomass, Elsevier, London
75. Thiele JH, Zeikus JG (1988) Appl. Environ, Microbiol. 52: 269
76. Daniels L, Fuchs G, Thauer RK, Zeikus JG (1977) J. Bacteriol. 132: 118
77. Heine-Dobbernack E, Choberth SM, Sahm H (1988) Appl. Environ. Microbiol. 54: 454
78. Tzeng SF, Bryant MP, Wolfe RS (1975) J. Bacteriol. 121: 154
79. Thauer RK, Jungermann K, Decker K (1977) Bacteriol. Rev. 41: 100
80. McInerney M, Bryant M (1980) Basic principles of bioconversion in anaerobic digestion and methanogenesis. In: Sofer S, Zaborsky O (eds) Biomass conversion processes for energy and fuels, Plenum. New York p 277
81. Cohen A (1982) Ph. D. Thesis, University of Amsterdam. The Netherlands.
82. Zoetemeyer RJ (1982) Acidogenesis of soluble carbohydrate containing wastewaters. Ph. D. Thesis. University of Amsterdam. The Netherlands.
83. McCarty R (1964)·Anaerobic waste treatment fundamentals. Public works 95, Sept.: 107, Oct.: 123, Nov.: 91
84. Niewenhoff FF (1981) Anaerobic digestion of dairy waste water. Second International Symposium on Anaerobic digestion, Travemunde. Germany. Elsewier/North Holland Biomedical press. The Netherlands, p 52
85. Pipyn P, Verstraete W (1979) Biotechnol. Lett. 1: 495
86. Ghosh S, Henry M (1982) Application of packed-bed upflow towers in two phase anaerobic digestion. First International Conference on Fixed-Film Biological processes. Kings Island, Ohio, April 1982

87. Boone DR, Bryant MP (1980) Appl. Envir. Microbiol. 40: 509
88. Price EC, Cheremissinoff (1981) Biogas production and utilization. Ann Arbor Science Publ. Michigan
89. Boyle WC (1976) Energy recovery from sanitary lanfields a review. In: Schegel Barnea (eds) Microbial Energy Conversion, Göttingen, p 119
90. Vesiland P (1975) Treatment and disposal of wastewater sludges. Ann Arbor, MI, p 30
91. Schroepfer GJ, Fuller WJ, Johnson AS, Ziemke NR, Anderson JJ (1955) Sewage Ind. Wastes 27: 460
92. Schroepfer GJ, Ziemke NR (1959) Sewage Ind. Wastes 31: 164
93. Callander IJ, Barford JP (1983) Process Biochemistry 18: 24
94. Martenson L, Frostell B (1982) Anaerobic waste water treatment in a carrier assisted sludge bed reactor. Proc. of the IAWPR Seminar on Aerobic Treatment of Waste Watr in Fixed Film reactor 16–18 June, Copenhagen, Denmark
95. Rands MB, Cooper DE (1966) Proc. 21st Ind. Waste Conf. Purdue Univ. p 613
96. Huss L (1982) In: Hughes DE (ed) Anaerobic digestion 1981, Elsevier, Amsterdam
97. Dague RR, McKinney RE, Pfeffer JT (1970) J. Wat. Pollut. Control Fed. 42: R29
98. Young JC. McCarty PL (1967) The anaerobic filter for waste treatment. Proc. of the 22nd Purdue Ind. Waste Conf. p 555
99. Young JC, McCarty PL (1969) J. Wat. Pollut. Control Fed. 41: R160
100. El Shafie AT, Bloodgood DE (1973) J. Wat Pollut. Control Fed. 45: 2345
101. Mueller JA, Mancini JL (1975) Proc. 30th Ind. Waste Conf. Purdue Univ. p 423
102. Braun R, Huss S (1982) Water Res. 16: 1167
103. Witt ER, Humphrey WJ, Roberts TE (1979) Proc. 34th Ind. Waste Conf. Purdue Univ. p 229
104. Suidan MT, Siekerka GL, Kao SW, Pfeffer JT (1983) Biotechnol. Bioengn. 25: 1581
105. Nyns EJ (1986) Biomethanation processes. In: Schönberg W (ed) Biotechnology, vol 8, chap 5, VCH, Weinheim
106. Young JC, Dahab MF (1982) Effect of media design on the performance of fixed bed anaerobic reactors. Proc. of IAWPR Seminar on Anaerobic Treatment of Waste Water in Fixed Film Reactors 16–18 June Copenhagen, Denmark
107. van den Berg L (1984) Can. J. Microbiol. 30: 975
108. van den Berg L, Kennedy KJ (1982) J. Chem. Technol. Biotechnol. 32: 427
109. van den Berg L, Kennedy KJ (1982) Effect of substrate composition on methane production rates of downflow stationary fixed film reactors. Proc. of Inst. of Gas Technology Symp. on Energy from Biomass and wastes. Lake Buena Vista FL. Chicago USA p 401
110. Hemens J, Meiring PG, Stander G (1962) Wat. Waste Treat. 9: 16
111. Cillie G, Henzen M, Stander G, Baillie R (1969) Wat. Res. 3: 623
112. Lettinga G, van Velson SW, Hobma W, de Zeeuw W, Klapwijk A (1980) Biotechnol. Bioeng. 22: 699
113. Dubourguier HC, Buisson MN, Tissier JP, Prensier G, Albagnac G (1987) Structural characteristic and methabolic activities of granular methanogenic sludge on a mixed defined substrate. Proc. of the GASMAT-Workshop, 25–27 October, Lunteren, The Netherlands, p 78
114. Lettinga G, Hulshoff-Pol LW, Hobma SW, Grin P, de Jong P, Roersma R, Ijspert P (1983) The use of a floating settling granular sludge bed reactor in anaerobic treatment. Proc. of the Eur. Symp. Anaerobic Waste Water treatment, 25–27 Nov., Noordwijkerhout, The Netherlands p 411
115. Kosaric N, Blaszczyk R and Orphan L Unpublished data.
116. De Zeeuw W, Lettinga C (1980) Acclimation of digested sewage sludge during start-up an upflow anaerobic sludge blanket (UASB) reactor. Proc. of the 35th Ind. Waste Conf. Purdue Univ. p 39
117. Lettinga G, Hobma SW, Hulshoff Pol LW, De Zeeuw W, de Jong P, de Grin P, Roersma R (1985) Water Science and Technology 15: 177
118. Lettinga G, Hobma SW, Hulshoff Pol LW, de Zeeuw W, Grin P, Roersma R (1982) Design, operation and economy of anaerobic treatment. Proc. of the IAWPR Seminar

on Anaerobic Treatment of Waste Water in Fixed Film Reactors 16–18 June Copenhagen, Denmark
119. Hulshoff Pol LW, de Zeeuw WJ, Velzeboer CTM, Lettinga G (1982) Granulation in UASB reactors. Proc. of IAWPR Seminar on Anaerobic Treatment of Waste Water in Fixed Film Reactors 16–18 June, Copenhagen, Denmark
120. Beeftink HH, Staugard P (1983) Acidification of glucose: architecture of biofilm as developed in an anaerobic gas-lift reactor with sands as adhesion support. Proc. Eur. Symp. Anaerobic Waste water Treatment. 23–25 Nov. Noordwijkerhort, The Netherlands p 107
121. Gorris LGM, van Deursen JMA (1987) Biofilm development in labscale methanogenic fluidized bed reactors. Proc. GASMAT-Workshop, 25—27 Oct. Lunteren, The Netherlands, p 87
122. Bonastre N, Paris JM (1988) Colonization and stimulation/inhibition properties of different support used in anaerobic fixed film reactors. Proc. of Fifth Int. Symp. on Anaerobic Digestion, 22–26 May, Bologna, Italy, p 11
123. Rouxhet PG, Mozes N (1987) Phisico-chemical bases of microbial adhesion. In: Ferranti MP, Ferro GL, L'Hermite P (eds) Anaerobic digestion: Results of research and demonstroation projects, Elsevier, London, p 236
124. Murray WD, van den Berg L (1981) J. Appl. Bacteriol. 51: 257
125. Verrier D, Mortier B, Dubourguier HC, Albagnac G (1988) Adhesion of anaerobic bacteria to inert support and development of methanogenic biofilms. Proc. of the Fifth Int. Symp. on Anaerobic Digestion, 22–26 May, Bologna, Italy p 61
126. Wheatley AD (1981) Envir. Technol. Lett. 2: 419
127. Heijnen JJ, Gist-Brocades NV (1983) Anaerobic wastewater treatment, Proc. Eur. Symp. Anaerobic Waste Water Treatment, 23–25 Nov. Noordwijkerhout, The Netherlands, p 259
128. Heijnen JJ (1982) Development of a high rate fluidized bed biogas reactor, 2nd Conference on Energy from Biomass, September, Berlin, West Germany
129. Jewell WJ, Switzenbaum HS, Morris JW (1981) J. Wat. Pollut. Control fed. 53: 482
130. Shieh WK, Keenan JD (1986) Fluidized bed biofilm reactor for wastewater treatment. In: Fiechter A (ed) Advances in Biochemical Engineering/Biotechnology vol 33 p 132, Springer, Berlin Heidelberg New York
131. Heijnen JJ (1983) Development of a high rate fluidized bed biogas reactor. In: Proc. Eur. Symp., 23–25 Nov. Noordwijkerhout, The Netherlands, p 259
132. Costerton JW, Geesey GG and Cheng KJ (1978) Sci. Am. 236: 86
133. Robinson RW (1986) Appl. Environ. Microbiol. 52: 17
134. Calleja GB (1984) Microbial aggregation, CRC Press, Boca Raton, FL
135. Robinson RW, Aldrich S, Hunt F, Bleiweis AS (1985) Appl. Environ. Microbiol. 49: 321
136. Zhylina TN (1976) Microbiologija 45: 414
137. Logan BE, Hunt JR (1986) Biotechnol. Bioengn. 31: 91
138. Kjelleberg S, Hermansson M (1984) Appl. Environ. Microbiol. 48: 497
139. Kosaric N et al. (1989) The University of Western Ontario, London, Ontario, Canada — unpublished data
140. de Vocht M, van Meenean P, van Assche P, Verstraete W (1983) Process Biochem. 18: 31
141. Switzenbaum MS, Robins JP, Hickey RF (1987) Immobilization of anaerobic biomass. Proc. of the GASMAT-Workshop, 25–27 Oct. Lunteren, The Netherlands, p 153
142. Huysman P, van Meenen P, van Assche P, Verstraete W (1983) Factors affecting the colonization of nonporous and porous material in model upflow methane reactor. 23–25 Nov. Noorwijkerhout, The Netherlands, p 187
143. Mozes N, Marchal F, Hermesse MP, van Haecht JL, Reubiaux L, Leonard AJ, Rouxhet PG (1987) Biotechnol. Bioengn. 30: 439
144. van Loosdrecht MCM, Lyklema J, Norde W, Schraa G, Zehnder AJB (1987) Appl. Environ. Microbiol. 531: 1898
145. Basu AK, Leclerc J (1975) Water Res. 9: 103
146. Ho CS (1986) Process Biochem. 21: 148

147. Dolfing J, Bloemen W (1983) Characterization of granular methanogenic sludge grown in upflow anaerobic sludge blanket (UASB) reactors. Proc. Eur. Symp. Anaerobic Waste Water Treatment, 25–27 Nov. Noordwijkerhout, The Netherlands, p 37
148. Grotenhuis JTC, Koorneef E, Plugge CM (1987) Immobilization of anaerobic bacteria in methanogenic aggregates. Proc. GASMAT-Workshop, 25–27 Oct. Lunteren, The Netherlands, p 52
149. Sam-Soon PALNS, Loewenthal RE, Dold PL, Marais GR (1988) Pelletization in upflow anaerobic sludge blanket reactors. Proc. of the 5th Int. Symp. on Anaerobic Digestion, Bologna, Italy, p 55
150. Sato T, Ose Y (1980) Water Res. 14: 333
151. Sheintuch M, Lev O, Einav P, Rubin E (1986) Biotechnol. Bioengn. 28: 1564
152. Forster CF, Rockey JS, Wase DAJ, Goswin SJ (1982) Biotechnol. Lett. 4(12): 799
153. Rudd T, Sterrit RM, Lester JN (1983) Biotechnol. Lett. 5(5): 327
154. Gijot S (1989) Biotechnology Research Institute, Montreal — private communication
155. Absolom DR, Lamberti FV, Policova Z, Zingg W, van Oss CJ, Neumann AW (1983) Appl. Environ. Microbiol. 46: 90
156. Wheatley AD (1981) Envir. Technol. Lett. 2: 419
157. Alibhai KRK, Forster CF (1986) Enzyme Microb. Technol. 8(10): 601
158. Robinson RH, Akin DE, Nordstedt RA, Thomas MV, Aldrich HC (1984) Appl. Environ. Microbiol. 48: 127
159. Lane AG (1986) Environ. Technol. Lett. 7: 555
160. Wiegant WM, de Man AWA (1982) Biotechnol. Bioengn. 28: 718
161. Bochem MP, Schoberth SM, Sprey B, Wengler P (1982) Can. J. Microbiol. 28: 500
162. Beeftink HH, Staugaard P (1986) Appl. Environ. Microbiol. 52: 1139
163. Wiegant JM (1987) The "spaghetti theory" on anaerobic granular sludge formation or the inevitability of granulation. Proc. GASMAT-Workshop, 25–27 Oct. Lunteren, The Netherlands, p 146
164. de Zeeuw W (1987) Granulation sludge in UASB reactors. Proc. GASMAT-Workshop, 25–27 Oct. Lunteren, The Netherlands, p 132
165. Salkinoja-salonen MS, Nyns EJ, Sutton PM, van den Berg L, Wheatley AD (1983) Wat. Sci. Tech. 15: 305
166. Kosaric N, Blaszczyk R, Orphan L (1990) Wat. Res. (in press)
167. Kosaric N, Blaszczyk R, Orphan L, Valladares J, Ngcakani Z, Lynch C, Richard A (1988) Anaerobic sludge granulation process. Annual Report. University of Western Ontario, London, Ontario, Canada
168. Eighmy TT, Maratea D and Bishop PL (1983) Appl. Environ. Microbiol. 45: 1921
169. Dubourguier HC, Prensier G, Albagnac G (1987) Structure and microbial activities of granular anaerobic sludge. Proc. GASMAT-Workshop, 25–27 Oct. Lunteren, The Netherlands, p 18
170. Thiele JH, Chartrain M, Zeikus JG (1988) Appl. Environ. Microbiol. 54: 10
171. Kosaric N, Mahoney EM, Varangu LK, Cairns WL (1987) Water Pollur. Res. J. Can. 22: 289
172. Valin S (1982) Predicting bioflocculation behaviour. Proc. of the 17th Can. Symp. on Water Pollution Research CCIW, Burlington, Canada, p 9
173. de Zeeuw W (1981) Use of anaerobic digestion for watewarer treatment. Antoine van Loeuwenhoek 46, p 110
174. Mahoney EM, Varangu LK, Cairns WL, Kosaric N, Murray RGE (1987). Wat. Sci. Tech. 19: 249
175. Hubbe MA (1981) Progress in Surface Science, 11: 137
176. Hulshoff Pol LW, Keijenkamp K and Lettinga G (1987) The selection pressure as a driving force behind the granulation of anaerobic sludge. Proc. of the GASMAT-Workshop, 25–27 Oct. Lunteren, The Netherlands, p 153
177. de Zeeuw WJ (1084) Acclimation of anaerobic sludge for UASB reactor start-up. Ph. D.-thesis. Agricultural University, Wageningen, The Netherlands

178. Wiegant WM, Claassen JA, Lettinga G (1985) Biotechnol. Bioengn. 27: 1374
179. Ono H (1965) Discussion on the utilization of material derived from treatment of wastes from molasses distillers by TR Bhaskaran. In: Baars J (ed) Advances in water pollution research, Pergamon, London, vol 2, p 102
180. Boone DR, Xun L (1987) Appl. Environ. Microbiol. 53: 1589
181. Brummel TE, Hulshoff Pol LW, Dolfing J, Lettinga G, Zehnder AJB (1985) Appl. Environ. Microbiol. 49: 1472
182. Klapwijk A, Smit H, Moore A (1981) In: Cooper PF, Atkinson (eds) Biological fluidized bed treatment of water and wastewater, Ellis Horwood, Chechester, p 205
183. van der Hoek JP (1987) Granulation of denitrifying sludge. Proc. of the GASMAT-Workshop, 25–27 Oct. Lunteren, The Netherlands, p 203
184. Pette KC, de Vletter R, Wind E, van Gils W (1981) Full-scale anaerobic treatment of beet-sugar wastewater. Proc. of the 35th Ind. Waste Conf. Purdue Univ. p 635
185. Trudell M, van den Berg L, Kosaric N (1985) Water Pollut. Res. J. Can. 20: 25
186. Kosaric N, Blaszczyk R and Orphan L (1990) Wat. Pollut. Res. J. of Canada, in Press
187. Blaszczyk et al. (1989) University of Western Ontario, London, Ontario, Canada — unpublished data.
188. Tilche A, Yang X (1987) Light and scanning electron microscope observation on the granule biomass of experimental SBAF and HABR reactors. Proc. of the GASMAT-Workshop, 25–27 Oct. Lunteren, The Netherlands p 170
189. Beeftink HH, van den Heuvel JC (1987) Biotechnol. Bioengn. 30: 233
190. Orphan L, Kosaric N, Blaszczyk R (1989) The form and composition of granules from UASB reactors. Proc. of the 7th Canadian Bioenergy R&D Seminar, 24–26 April, Ottawa, Canada
191. Lettinga G, Hufshoff Pol LW, Grin P, de Jong P, Roersma R, Ijspert P (1983) The use of a floating settling granular sludge bed reactor in anaerobic treatment. Proc. of the Europ. Symp. Anaerobic Waste Water Treatment, 23–25 Nov. Noorwijkerhout, The Netherlands, p 411
192. Li DH, Garnaczarczyk Y (1987) Wat. Res. 21: 257
193. Tramper J, van Groenestijn JW, Luyben KChAN, Hulshoff Pol (1984) Innovation in Biotechnology, p 145
194. Iza J, Garcia PA, Sanz I, Fdz-Polanco F (1987) Granulation results in anaerobic fluidized bed reactors. Proc. of the GASMAT-Workshop, 25–27 Oct. Lunteren, the Netherlands, p 195
195. Kosaric et al. (1989) University of Western Ontario, London, Ontario, Canada — unpublished data
196. Guiot SR, Pauss AP, Bourque D, Houseini HE, Lavioe L, Bealieu C, Samson R (1988) Effect of upflow liquid velocity on granule size distribution in an upflow anaerobic bed-filter (UBF) reactor. Fifth Int. Symp. on Anaerobic Digestion. Bologna, Italy, p 121
197. Cristensen DR, Gerick J, Eblen JE (1984) J. Wat. Pollut. Control Fed. 56: 1059
198. de Zeeuw W (1987) Granular sludge in UASB reactors. Proc. of the GASMAT-Workshop, 25–27 Oct. Lunteren, The Netherlands, p 133
199. van Bellegem TI (1980) Biotechnol. Lett. 2: 219
200. Joseph Oat Corporation. The Oat CSM upflow anaerobic sludge blanket process — summary of pertinent facts, Camden, New Jersey, USA, 08104
201. Samson R, van den Berg B, Peters R, Hade C (1984) Dairy waste treatment using industrial-scale fixed film and upflow sludge bed anaerobic digester: design and start-up experience. Proc. of the 38th Ind. Waste Conf. Purdue Univ. p 235
202. Food processing and beverage industries anaerobic wastewater treatment. Technical document. Paques-Lavalin (BIOPAQ), Toronto, Canada
203. Lettinga G, Hulshoff Pol L (1986) Wat. Sci. Tech. 18: 99
204. Maat DZ, Habets LH (1987) Anaerobic treatment: a low cost and reliable wastewater treatment process. In: Food processing and beverage industries anaerobic wastewater treatment. Paques-Lavalin (BIOPAQ), Toronto, Canada

205. Kosaric et al. University of Western Ontario, London, Canada — unpublished data
206. Cohen A, Breure AM, van Andel JG, van Deursen A (1980) Wat. Res. 14: 1439
207. Cohen W, Breure AM, van Andel JG, van Deursen A (1982) Wat. Res. 1: 449
208. Breure AM, Beeftink HH, Verkuylen J van Andel JG (1986) Appl. Microbiol. Biotechnol.
     23: 295

Production of Ethanol from Lignocellulosic Materials: State of the Art

L. Vallander and K.-E. L. Eriksson*
STFI, Box 5604, S-11486 Stockholm, Sweden

Production of ethanol from sugar (hexoses) and starch is carried out by a well developed technology. Production of ethanol from lignocellulosic materials is more difficult, requires several process steps, and is as yet in the development stage.

If the polysaccharides in the lignocellulosic materials are to be saccharified via enzymic hydrolysis, the raw material must be pretreated if a high yield of sugars is to be expected. Pretreatment of the raw material with steam at high temperature and pressure is the preferred method. With this process substantial parts of the lignin and the hemicelluloses are degraded to products extractable with water, ethanol or alkaline solutions. Even if further development of this pretreatment technique may be necessary it is clear that the already existing technique can be used on a commercial scale.

The enzymic hydrolysis of complicated solid substrates as are the lignocellulosic materials is a slow process. A reduction of the time necessary to achieve satisfactory sugar yields will therefore have a large impact on the process economy.

Several factors are of importance for the sugar yield in the enzymic hydrolysis. Particularly important are the composition of the enzyme mixture and the ratios between enzyme and substrate,

---

* Present address: University of Georgia, Department of Biochemistry, Athens, Georgia 30602, USA

Advances in Biochemical Engineering/
Biotechnology, Vol. 42
Managing Editor: A. Fiechter
© Springer-Verlag Berlin Heidelberg 1990

the inhibition of the enzymes by degradation products, the adsorption of enzymes onto the substrate and the possibility of reutilizing both enzymes in solution and enzymes adsorbed onto the substrate.

As mentioned earlier, commercial techniques are available for the bioconversion of hexoses to ethanol. A similar technique for the bioconversion of pentoses is still lacking. However, important progress is being made in this field.

The interest in developing commercial processes where lignocellulosic materials are the basis for production of ethanol will always be strongly dependent upon the oil price and the availability of oil, even though the environmental aspects on fuel most likely will be more important in the future. It is our strong belief that sooner or later a process for the conversion of lignocellulosics to ethanol will be in demand.

1 Introduction

The production of ethanol from carbohydrates can be a more or less difficult task, depending on the choice of raw materials. The use of starch or monomeric hexoses as substrates presents no problems since well-developed technologies have long existed [1]. The production of ethanol and other useful products from lignocellulosic materials by processes involving biotechnological steps, i.e. enzymic hydrolysis as the hydrolysis step, is however a more complex undertaking. The nature of the substrate then makes it necessary to include more steps in the process (Fig. 1).

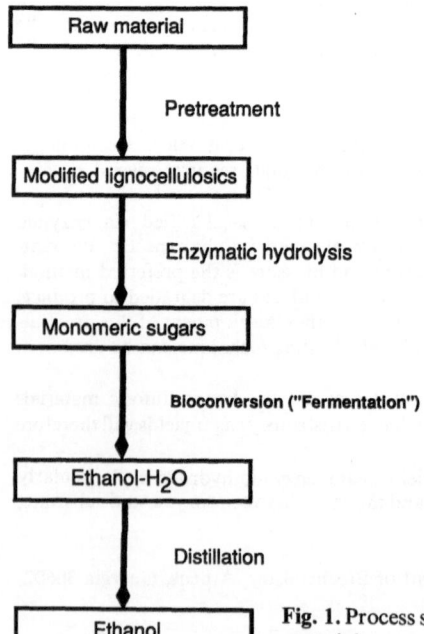

Fig. 1. Process steps in the production of ethanol from lignocellulosic materials

The lignocellulose substrate has to be pretreated to make it as susceptible as possible to the action of the enzymes. This can be achieved by several different methods. In principle, the pretreatment should cause disintegration of the material, thereby creating a large surface area on which the enzymes can work. Increased accessibility can also be achieved by solubilizing the lignin or the hemicelluloses. A washing or extraction step is therefore often inclued in the pretreatment to separate soluble components from the polymeric substrate.

After pretreatment, the substrate is ready for the enzymic hydrolysis. In principle there are two ways to organize such a process. A sugar solution is produced separately in a first step followed by a bioconversion ("fermentation") to ethanol in a second step, or the hydrolysis and "fermentation" steps are combined in a single step. The advantage of the first route is that a higher hydrolysis temperature can be used, which increases the rate of saccharification. On the other hand, combined hydrolysis and "fermentation" (CHF) reduces the catabolite repression of the cellulases to a minimum by keeping the sugar levels low.

The feasibility of several lignocellulosic materials for ethanol production has been studied around the world, depending on what is locally available. The main interest appears to be focused on aspen wood, poplar wood, different kinds of straw and sugar cane bagasse (see Table 1). Besides cellulose and lignin, the majority of the studied materials also contain acetylated xylans in the range of 10–25% [2, 3]. The main sugar products after enzymic hydrolysis are therefore glucose and xylose.

This review is divided into three main parts:

A. Pretreatment
B. Enzymic hydrolysis
C. Bioconversion ("fermentation")

and these are dealt with in Sects. 2, 3 and 4 respectively.

The present status of these areas can be summarized as follows:

Even if more work is justified to develop further the pretreatment techniques and their application to specific substrates, it is probably correct to say that efficient pretreatment methods already exist. Steam pretreatment has for example gained wide general acceptance as a highly efficient and economically feasible method. Pretreatment is therefore not a major obstacle to the conversion of lignocellulosic materials to ethanol.

The emphasis of this review is on enzymic hydrolysis. Acid hydrolysis is not included at all.

As we see it, the complexity of the enzymic hydrolysis process depends on the fact that it is a multivariate system. Success in developing efficient bioconversion processes will depend on our ability to unterstand the interrelationships between the different variables involved and their dynamics. Some of these variables are:

— the substrate (chemical) composition after pretreatment, lignin content, the degree of crystallinity of the cellulose and the extent of amorphous regions, the surface structure of the substrate, pore volume, pore size distribution);
— the changes in the substrate during hydrolysis (production of enzyme inhibitors);
— the composition of the enzyme mixture added;

— the enzyme adsorption profile during the process;
— the relationship between enzyme and substrate concentrations;
— the recirculation of enzymes.

Enzymic hydrolysis of a solid substrate is a slow process. Any reduction in the time needed to obtain a satisfactory sugar yield will therefore have a significant impact on the process economics. The key word here is "dynamics". Due to the multivariate character of the process we can expect that the rate-determining factor will change with time.

The main product of the hydrolysis is glucose. The technology for the bioconversion of glucose and other hexoses is well-known. Pentose (i.e. xylose) bioconversion is less developed technically but intense research during the last decade has resulted in significant advances in this field.

2 Part A: Pretreatment

2.1 Evaluation of Pretreatment Techniques

The intention of the pretreatment is to open the structure of the lignocellulosic material, thereby making it accessible to the cellulolytic enzymes. This is accomplished in different ways, e.g. by increasing the surface area, by removing lignin or by solubilizing hemicelluloses. The pretreatments are generally divided into mechanical, chemical, biological, physical and combined techniques. Table 1 summarizes data from studies on pretreatment.

Pretreatments in one group have in common that water (as liquid or steam) is used at elevated temperatures. Depending on the pretreatment conditions, they have been given different names.

"Autohydrolysis" — alone or together with steam explosion — is the name given to steam pretreatment in the approximate temperature range of 170 to 200 °C [8].

"Steam explosion" is the most widely used term. The temperature range is here extended to 250 °C, and the pretreatment is ended by a sudden release of pressure.

The term "steam pretreatment" is advocated by Brownell and Saddler [11, 12], due to their finding that the explosion contributes little, if at all, to the enzymic hydrolysability of the pretreated substrate. All these names will appear in the text, but the term "steam pretreatment" is used in preference. "Hydrothermolysis" is a pretreatment in which the raw material is subjected to liquid water at high temperature and pressure but where no steam appears [7].

Rao et al. [4] studied two alkaline pretreatments a) 1 M NaOH + washing and b) 1 M NaOH + neutralization of sugar cane bagasse, and compared them with steam explosion. They reported that steam explosion or alkali pretreatment + neutralization in situ resulted in 59 % and 63 % conversion of bagasse to sugar in 48 h of enzymic hydrolysis, respectively. The enzyme dosage was high, 100 FPU (Filter Paper Units), per g of substrate. When the enzyme load was reduced to

Table 1. Pretreatment and enzymic saccharification of different lignocellulosic materials

Pretreatment	Substrate	Enzyme		Time h	Saccharification[b]		Ref.
		Source	Conc. FPU* g^{-1} substr.		Pretr. mtrl. %	Original mtrl. %	
1N NaOH	Bagasse	P.f.	25	24	63	34	[4]
	Bagasse	P.f.	25	48	70	37	
Steam expl.	Bagasse	P.f.	25	24	45	45	
	Bagasse	P.f.	25	48	52	52	
Untreated	Bagasse	P.f.	25	24	—	15	
	Bagasse	P.f.	25	48	—	17	
	Cotton	P.f.	50	21 d	—	97	
Steam expl.	Wheat straw	T.r.	100	24	75	—	[5]
H$_2$O$_2$/OH	Wheat straw	T.r.	100	24	56	—	
Defibration	Wheat straw	T.r.	100	24	29	—	
Untreated	Wheat straw	T.r.	100	24	—	10	
Steam expl.	Aspen	T.r. C30	2200	2	70	—	[6]
	Aspen	QM 9414	2250	2	80	—	
	Aspen	T. E58	1750	2	75[c]	—	
HNO$_3$	Aspen	T.r. C30	2200	2	30	—	
	Aspen	QM 9414	2250	2	27	—	
	Aspen	T. E58	1750	2	34	—	
Hydrothermo-lysis	Poplar	T. viride	1.75	70	75	—	[7]
	Wheat straw	T. viride	1.75	70	70	—	
Organosolv	Poplar	T. viride	1.75	70	90	—	
	Wheat straw	T. viride	1.75 -	70	80	—	
Autohydrolysis	Bagasse	T.r. C30	20	24	51	—	[8]
	Bagasse	T.r. C30	20	48	70	—	
	Bagasse	T.r. C30	20	72	83	—	
Untreated	Bagasse	T.r. C30	20	24	—	14	
	Bagasse	T.r. C30	20	48	—	18	
	Bagasse	T.r. C30	20	72	—	19	
NaOH	Bagasse	Onozuka		24	65	—	[9]
NH$_3$	Bagasse	Onozuka		24	25	—	
SO$_2$	Bagasse	Onozuka		24	45	—	
Steam expl.	Bagasse	Onozuka		24	85	—	
Organosolv (MeOH)	Bagasse	Onozuka		24	73	—	
SO$_2$ impreg. steam expl.	P. radiata	T.r.	20	72	—	80–84	[10]

Abbreviations: P.f. = *Penicillium funiculosum*

T.r. = *Trichoderma reesei*

[a] FPU = Filter Paper Units;

[b] The calculation of saccharification has been based on the carbohydrate content of pretreated and original materials, respectively;

[c] Reducing sugar in percent of total steam exploded material.

25 FPU g^{-1}, which is a more suitable level on a technical scale, the saccharification dropped by 7–9%. However, in the opinion of Rao et al. [4] steam explosion is the best choice, since the cost and environmental impact of the alkali pretreatment makes it less attractive.

Vallander and Eriksson [5] studied the effects of steam explosion, alkaline hydrogen peroxide pretreatment (5% H_2O_2 on wheat straw, 1 h, pH 10.0—12.3) and defibration of wheat straw, respectively, on the yields of reducing sugar obtained in enzymic hydrolysis. The first two methods resulted in an efficient saccharification within 24 h of enzymic hydrolysis, steam explosion being the best.

Saddler et al. [6] using aspen wood as substrate, studied nitric acid pretreatment, steam explosion and multiple pretreatments starting with steam explosion. The subsequent pretreatments were (water + alkali) extraction + drying, extraction + drying + Wiley milling, or only chlorite pretreatment. Steam explosion (250 °C, 20–80 s) created a more easily hydrolysable substrate than that obtained with a 1–3 h nitric acid pretreatment. Drying the steam-treated aspen wood had a negative effect on enzyme accessibility, particularly on those samples which had shown the greatest hydrolysis prior to drying. Chlorite treatment of steam-exploded wood gave rise to an increased enzymic hydrolysis, most likely due to removal of lignin. If the dried samples were milled in a Wiley mill, a positive effect on enzymic hydrolysis was observed particularly with the samples showing the lowest degree of hydrolysis after drying. On the other hand, it was observed that milling resulted in a loss of enzyme accessibility in the pretreatment sequence: steam explosion — extraction — drying — milling, when the steam-exploded extracted substrate had a high accessibility.

Bonn et al. [7] compared the influence of hydrothermal and organosolv pretreatment of poplar wood and wheat straw on enzymic hydrolysis. Hydrothermolysis was more suitable as pretreatment of wheat straw while organosolv pretreatment resulted in a better enzymic saccharification of poplar wood. Both pretreatment methods were efficient, giving glucose yields of 80–90% of the theoretical values in 70 h of enzymic hydrolysis.

Dekker and Wallis [8] reported that autohydrolysis-steam explosion of bagasse produces a substrate where the cellulose is rapidly saccharified enzymatically. Extraction of the steam-exploded material with alkali or ethanol to remove lignin lowers the degree of saccharification considerably. The autohydrolysis-steam explosion was as effective in promoting hydrolysis as was 0.25 M NaOH (70 °C, 2 h).

Rolz et al. [9] studied pretreatment and enzymic hydrolysis of sugar cane chips using NaOH, Na_2CO_3 + CaOH, SO_2, steam explosion, alkaline organosolv, aqueous phenol and a full soda cook as pretreatments. Steam explosion solubilized the hemicelluloses completely and it also promoted the greatest enzymic saccharification.

Hardwoods and other materials containing high amounts of acetylated xylans have long been successfully steam pretreated and enzymatically saccharified. Mamers and Menz [13] and recently also Clark and Mackie [10] have shown that softwoods can be efficiently hydrolyzed enzymatically if they are impregnated with SO_2 prior to steam treatment. With such a pretreatment of *Pinus radiata*, 82% saccharification of the cellulose to glucose was obtained in 72 h.

Morjanoff and Gray [14] achieved an improved hydrolysability of steam-exploded sugar cane bagasse if 1 g H_2SO_4 100 g^{-1} dry bagasse was added prior to the steam treatment. A sugar yield of 83–84% calculated from the polysaccharide content of untreated bagasse was obtained in 40 h of enzymic hydrolysis.

The wide variety of substrates and pretreatments listed in Table 1 gives a unanimous picture. Steam pretreatment (steam explosion, autohydrolysis) converts a lignocellulosic raw material to a substrate well suited for enzymic hydrolysis.

It should, however, be observed that xylan solubilized by steam pretreatment gives rise mainly to obligosaccharides and only to relatively minor amounts of the xylose monomer [15, 16]. Some kind of treatment thus seems to be necessary to convert the xylo-oligomers to the bioconvertible xylose.

A pretreatment method related to steam explosion is the freeze-explosion technique proposed by Dale and Moreira [17], which is based on the treatment of the cellulosic material with a volatile liquid under pressure followed by pressure release to evaporate the liquid and to reduce the temperature. A saccharification yield greater than 90% was achieved by enzymatic hydrolysis of alfalfa and rice straw after pretreatment with liquid ammonia. Volatile liquids such as ammonia, which are able to swell and decrystallize the cellulose were particularly efficient in this pretreatment.

The term "steam explosion" which has long been used, may eventually give way to "steam pretreatment" as suggested by Brownell and Saddler [11, 12]. The chemical reactions promoted by the high temperature and steam pressure appear to be the main causes of the increased enzymic saccharification.

A high degree of enzymic digestibility, comparable to the effect of steam pretreatment, can also be achieved by chemical pretreatment, although the chemical pretreatment often has a definite drawback in that an effluent problem is created. More expensive materials in the process equipment may also be needed.

Morphological differences between substrates clearly affect the rate of hydolysis. Grethlein [18] has shown that the initial rate of saccharification (during the first 2 h) is proportional to the total volume of pores large enough to allow penetration by the cellulase molecules (Fig. 2). The lower rate of hydrolysis of pine over hardwood species could be explained by the smaller number of penetrable pores in the former. It was also shown that with the chosen (acid) pretreatment, the number of large pores increased many times, thus improving the accessibility to the enzymes.

It has also been shown by Grous and coworkers [19] that steam explosion increases the pore volume accessible to the enzyme molecules, whereas drying of the substrate, which is known to affect its hydrolysis, resulted in a reduction in the pore volume. Similarly, Wong et al. [20] reported that steam explosion increases cellulose digestibility by increasing fiber porosity, i.e. pore volume accessible to the enzymes.

Chen and Grethlein [21] have recently studied the effect on saccharification of reducing the size of the cellulase molecules. The size of the enzyme was reduced by protease treatment without decreasing the enzyme activity. A minor increase in the endo-glucanase activity towards the solid substrate was observed. Hydrolysis of hardwood was, however, less effective with the pretreated enzymes. The decline in hydrolysis yield correlated well with the reduction in size of the protease-treated cellulase. The reduction in enzyme size appears to reduce the size

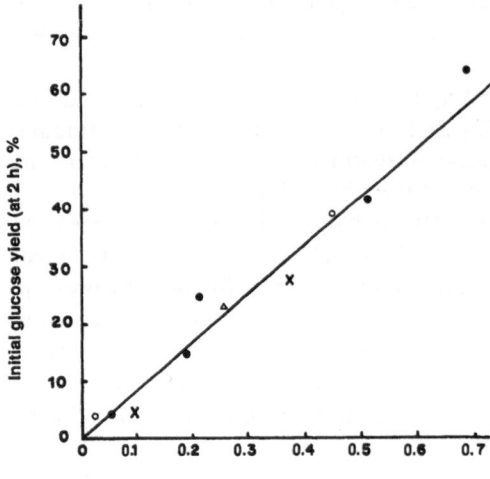

Fig. 2. Initial glucose yield (at 2 h) as a function of pore volume (ml g^{-1}) accessible to the solute with a nominal diameter of 51 Å for various substrates: mixed hardwood, ●; poplar, ○; white pine, ×; and steam extracted pine, △. (Grethlein [18]). (c) 1985, BIO/TECHNOLOGY. Used by permission.

Pore volume, ml g^{-1}, available to 51 Å

of the binding area of the enzyme to the substrate. This observation is in agreement with the conclusion of Rabinovich et al. [22] that the hydrolytic capability of cellulases correlates positively with their ability to bind to the substrate. To conclude:

— Lignocellulosic materials have to be pretreated to make them sufficiently susceptible to enzymic hydrolysis. At best, a 20 % saccharification has been achieved without pretreatment, (Table 1).

— Steam pretreatment at high temperatures, 170 °C–250 °C (autohydrolysis, steam explosion) is an efficient method of preparing substrates for enzymic hydrolysis.

— The pretreatment conditions, whatever method is chosen, should be adjusted to the substrate used.

— A succession of different petreatments does not necessarily result in a better enzymic saccharification than a single pretreatment [6].

— The pretreatment improves saccharification by making the substrate more accessible to the enzymes. Hydrolysis is thus promoted by a reduction of particle size, [23, 24], but the rate of hydrolysis is particularly linked to the specific surface area [25] or pore volume accessible to the enzymes [18].

— Drying of a lignocellulosic substrate is detrimental to the enzymic hydrolysis efficiency [6, 26].

3 Part B: Enzymic Hydrolysis

3.1 Background

The generally accepted picture of the degradation of cellulose by most of the wood-rot fungi is that it proceeds by the synergistic action of three types of hydrolytic enzyme: *endo*-1,4-β-glucanase, *exo*-1,4-β-glucanase, and 1,4-β-glucosidase. The *endo*-glucanases cleave the cellulose molecules at random and generate nonreducing ends as substrate for the *exo*-glucanases (cellobiohydrolases). In contrast to *exo*-cellulases, *endo*-glucanases hydrolyse substituted celluloses such as carboxylmethylcellulose (CMC). The mode of action and the reason(s) for the strong interaction exhibited by fungal cellulases is not yet clearly understood [27].

The most extensively studied cellulases are those from *Trichoderma reesei*. At least four components of its cellulolytic system have been purified to apparent homogeneity. Amino acid sequences of some enzymes are known and full length cDNA copies of all genes from *T. reesei* coding for, respectively, CBH I, CBH II, EG I, and EG III have been isolated [28].

Elucidation of the primary structures of these four components of the *T. reesei* cellulase system revealed obvious homologies between CBH I and EG I ($> 45\%$). Perhaps the most striking feature emerging from the sequence analyses is the presence of short, conserved regions (AB blocks) at either the N-termini (CBH II, EG III) or C-termini (CBH I, EG I), [29]. Block A is 30 amino acids long and block B 40 amino acids long. Block B seems to serve as a hinge region linking block A to the respective core (active site) enzymes. The core proteins of CBH I and CBH II (50–55 kDa), obtained by partial proteolysis, retains their full catalytic activity against water-soluble substrates but their activity towards microcrystalline cellulose is almost completely (core I) or partially (core II) lost. Since the loss of the AB blocks leads to a decreased adsorption to cellulose, the conclusion is that these blocks are of importance in binding the enzyme to the cellulose substrate.

For the hydrolysis of cellulose only enzymes hydrolysing 1,4-β-glucosidic linkages are necessary. The structures of the hemicelluloses of wood are, however, more variable, involving linear 1,4-β-linked chains of xylose or mannose which, substituted with other sugars or acetyl groups, make up the branched heteropolysaccharides which constitute hemicelluloses. A more complex set of enzymes is therefore required for their degradation. Complete degradation of a branched, acetylated xylan requires the concerted action of several different hydrolytic enzymes, including *endo*-1,4-β-xylanase, 1,4-β-xylosidase, α-glucosidase, α-L-arabinofuranosidase and acetylxylanesterase. Considerable progress has recently been made in the separation and characterization of these enzymes [30].

The endoxylanases are the best characterised and most widely studied of the hemicellulolytic enzymes. These enzymes initiate an end-wise attack on the backbone of xylans to produce both substituted and non-substituted short-chain oligomers, xylobiose and xylose. To convert the water-soluble oligomers, dimers, etc. to xylose, β-xylosidases are employed. The enzymes releasing the substituents on the xylan backbone, i.e., α-L-arabinosidase, α-D-glucuronidase, and acetylxylanesterase, act in synergism with *endo*-xylanases and β-xylosidases.

The mannan hemicelluloses, *galacto*-glucomannans and glucomannans, are both branched heteropolysaccharides and the concerted action of several enzymes is again required for their complete hydrolysis. Enzyme preparations suitable for such hydrolysis require the concerted action of the following hydrolytic enzymes: *endo*-1,4-β-mannanase, 1,4-β-mannosidase, 1,4-β-glucosidase and α-galactosidase. Enzymic hydrolysis of mannans has recently been reviewed by Dekker [31].

The most widely studied and best characterized of the mannan degrading enzymes are the *endo*-mannanases. They attack the backbone of the mannans in an end-wise manner to produce shorter, substituted and non-substituted oligomers, mannobiose, and mannose. β-Mannosidase then convert the water-soluble oligomers and dimers to mannose [32]. The β-glucosidases and α-galacto-sidases, which release the glucose and galactose substituent on the mannan backbone, act in synergism with *endo*-mannanase and β-mannosidase.

3.2 Enzyme/Substrate Ratios and Composition of the Enzyme Mixture

The choice of substrate and enzyme concentrations in an enzymic hydrolysis process must be based on several economic considerations. Since the cost of the enzyme is high, a low enzyme dosage is preferred. A long hydrolysis time can compensate for a low enzyme addition, but this can lead to excessively high capital costs since a long hydrolysis time requires large volume equipment. A low substrate concentration gives a high saccharification yield, but the dilute sugar solution produced is expensive to concentrate in downstream processing. The cost of raw material and energy, the extent of enzyme recovery, the production scale and the value of the by-products are other factors of importance.

It thus appears that an optimal enzyme-to-substrate ratio exists for a given enzymic hydrolysis process depending upon the location of the factory and the particular set-up of the process. Little work has been done to elucidate the import-ance of the enzyme-to-substrate ratio.

As expected, the enzyme-to-substrate ratio (FPU per g of substrate) is not linear. Factors such as substrate accessibility and product inhibition influence the hydrolysis rate, so that a doubling of the enzyme dosage gives less than a doubling of the sugar yield within a given time. Vallander and Eriksson [5] reported a 35% saccharification yield of wheat straw in 24 h with an enzyme addition of 20 FPU g^{-1} substrate. No additional β-glucosidase was given in this experiment. In order to double the saccharification to 70%, an enzyme dosage of 100 FPU g^{-1} of wheat straw had to be used (Fig. 3).

Tan et al. [33] studied the hydrolysis of Solka Floc in a column reactor with extra addition of β-glucosidase (FPU: β-glucosidase U 1:1). A moderate increase in sugar concentration from 4.5 to 6.5% was observed when the enzyme concen-tration was tripled from 10 to 30 FPU g^{-1} of cellulose.

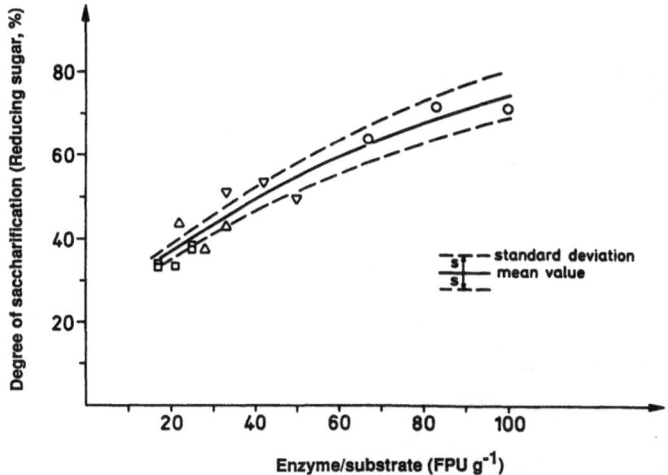

Fig. 3. The degree of saccharification (reducing sugar) as a function of the enzyme: substrate ratio. Concentration of wheat straw: (○) 6%, (▽) 12%, (△) 15.5%, (□) 18.3%. (Vallander and Eriksson [5]) Biotechnology and Bioengineering. Copyright (c) 1985, John Wiley & Sons, Inc. Reprinted by permission of John Wiley & Sons, Inc.

Autohydrolysed-steam exploded sugar cane bagasse was saccharified with *Trichoderma reesei* C-30 cellulase by Dekker and Wallis [8]. Without the addition of extra β-glucosidase, 20 FPU g^{-1} of substrate of this potent cellulase preparation had to be added to yield 70% of saccharification in 24 h, but the addition of β-glucosidase (FPU:β-glucosidase U = 1:1.25) resulted in 90% saccharification under otherwise identical experimental conditions. Dekker et al. [34] reported the saccharification of autohydrolysed *Eucalyptus regnans* sawdust. Saccharification of 40% of the cellulose was achieved in 24 hours by 20 FPU g^{-1} of substrate with no exogeneous β-glucosidase given (FPU:β-glucosidase U = = 1:0.29). With an added amount of β-glucosidases (FPU:β-glucosidase U = = 1:1.25) cellulose saccharification increased to 80% (Fig. 4).

A high sugar concentration can be obtained if the substrate is added successively. Perez et al. [35] reported that stepwise addition of corn stover to 25% concentration within 2.5 hours resulted in a sugar concentration of 9% after 24 hours of enzymic hydrolysis. For a given enzyme dosage, this technique will maintain a greater enzyme-to-substrate ratio during hydrolysis and consequently a higher rate of saccharification can be anticipated.

It appears that a high degree of saccharification can be achieved with no more cellulase added than 20 FPU g^{-1} of substrate, provided that the substratè has been properly pretreated, that an efficient cellulase preparation is used and that sufficient quantities of β-glucosidase are present during the hydrolysis.

The amount of enzyme needed to obtain a sufficient saccharification of a given substrate is clearly linked to the composition of the enzyme mixture. Data presented above show that an adequate level of β-glucosidase during hydrolysis is necessary. This is usually accomplished by the addition of β-glucosidase from *Aspergillus niger* or *A. phoenicis*.

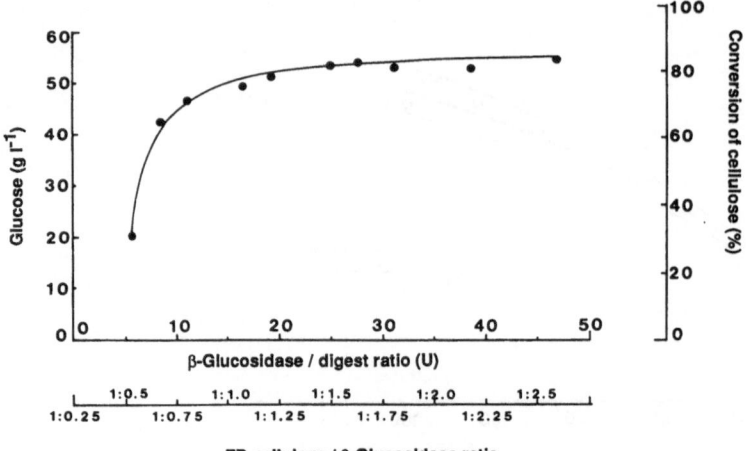

Fig. 4. The effect of adding exogeneous β-glucosidase on the enzymic saccharification of auto-hydrolysis-exploded *Eucalyptus regnans* sawdust. The enzymic digests contained pretreated sawdust (1g DW), cellulase (20 FPU), 50 mM acetate buffer (pH 5.0) and variable amounts of β-glucosidase in a final volume of 10 ml and were incubated for 24 h at 50 °C. (Dekker et al. [34]) Courtesy by HARWOOD ACADEMIC PUBLISHERS GmbH

Table 2

Substrate	Strain[a]	FPU β-glucosidase units^{-1}		Remark	Ref.
		Initial level	Level to remove cellobiose inhib.		
Bagasse	T.r. 30	1:0.57	1:1.0		Dekker and
	T.r. QM 9414	1:0.28			Wallis [8]
Eucalyptus regnans	T.r. C-30	1:0.29	1:1.0–1.25		Dekker et al. [34]
Spruce sulfite pulp	T.r. MCG 77	1:0.22			Esterbauer et al. [36]
Solka Floc BW 300	T.r. (Celluclast)	1:0.19	1:2.06		Tan et al. [33]
Solka Floc BW 200	T.r. QM 9414	1:0.12	1:1.0	CHF[b]	Ghosh et al. [37]
		1:0.97	(ca) 1:1	CHF	
			(ca) 1:5	Hydrolysis	
Solka Floc BW 200,					Sternberg et al. [38]
Avicel pH 102	T.r. QM 9414	1:0.25	1:0.25		

[a] T.r. = *Trichoderma reesei*;
[b] CHF = Combined hydrolysis and fermentation.

Table 2 presents data on the concentration of β-glucosidase in *T. reesei* cellulase preparations as reported in the literature and the β-glucosidase levels necessary to remove the cellobiose inhibition of the *endo-* and *exo-*glucanases.

For most *T. reesei* cellulases shown in Table 2 the original β-glucosidase activity amounts to 0.2–0.3 times the FPU. To attain a high degree of saccharification, the ratio of β-glucosidase units to FPU must be raised to at least 1:1 or somewhat higher. The differences observed may be due to the fact that different enzyme mixtures and substrates have been used.

Ghosh et al. [37], using a 1:1 FPU:β-glucosidase ratio during combined hydrolysis and "fermentation" (CHF), report that high levels of cellobiose were still present at FPU:β-glucosidase ratios of about 1:5 during hydrolysis. The reason for this has not been elucidated.

Esterbauer et al. [36] remark that a sufficiently high level of β-glucosidase can be obtained by choosing the correct growth conditions for the enzyme-producing organism. If such conditions are established, no addifion of exogeneous β-glucosidase is needed.

3.3 Competitive Inhibition of the Cellulases

Cellobiose is a strong inhibitor of both *endo-* and *exo-*glucanases while glucose mainly inhibits the β-glucosidase activity. Different ways have been suggested to cope with these problems. The most obvious way to alleviate the product inhibition caused by cellobiose is to add an exogeneous β-glucosidase.

Simultaneous saccharification and fermentation are proposed by Ghosh et al. [37] as one way to minimize competitive inhibition. This makes it possible to use a smaller amount of β-glucosidase than if the hydrolysis step is run separately. Compared to the other data presented in Table 2, very little β-glucosidase can be saved this way. Alternatively, inhibition by glucose can be relieved by its continuous removal by ultrafiltration. High sugar yields, > 90%, have been reported for this technique [39].

Khan et al. [40] used charcoal to minimize the inhibition by selectively adsorbing cellobiose and glucose. When 5 g of charcoal per g of cellulose was used for treatment of the hydrolysis mixture, an equal or improved saccharification was achieved compared to the addition of 60 U of β-glucosidase. However, handling huge amounts of charcoal on a technical scale and the introduction of an extra extraction step to separate and recover the sugars and the charcoal does not seem very attractive.

Fadda et al. [41] report that immobilized enzymes are more resistant to inhibition by cellobiose and glucose. The immobilized enzymes were also more stable than the soluble enzymes and solubilized three solid lignocellulosic substrates significantly more than soluble enzymes.

Ghosh et al. [37] observed inhibition of the cellulases by as little as 0.75% ethanol. The inhibition is, however, less severe than with cellobiose or glucose. Vallander and Eriksson [5] also found that cellulases were inhibited more by glucose than by ethanol.

Even when the sugar concentration in the hydrolysis liquor is very low, 1–2 %, inhibition of the cellulases is observed [42]. It appears that sugar adsorbed to solid substrate creates a microenvironment around the enzyme molecules which supresses their activity. Considerable amounts of sugar may indeed be present on the surface of the substrate [42]. Removal of the hydrolysate and resuspending the substrate (with enzymes adsorbed to it) in fresh liquid reduces inhibition and improves the degree of saccharification [5, 42].

3.4 Enzyme Adsorption

The adsorption properties on the substrates of the participating enzymes are vital for the rate and degree of saccharification and they have therefore received a great deal of attention. Lee et al. [25] studied the adsorption of cellulases on several different substrates. Two types of adsorption curves were obtained:
1) When physicochemically pretreated substrates are used, enzymes become rapidly and extensively adsorbed onto the substrate. This is due to a good accessibility and is followed by a continuous decrease in adsorption.
2) If the substrate has not been pretreated or contains materials difficult to hydrolyse, such as cotton, untreated Solka Floc or newsprint, the enzyme accessibility is low, i.e. the initial enzyme adsorption is slow but increases with time as the continuing hydrolysis makes the substrate more accessible.

Vallander and Eriksson [5] studied the course of enzyme adsorption during the first 24 h of hydrolysis. In agreement with Lee et al. [25], a rapid initial adsorption (70 % of 3 FPU ml^{-1} added) on steam-pretreated wheat straw was observed (Fig. 5). However, a limited desorption of enzymes occurs during the first hours

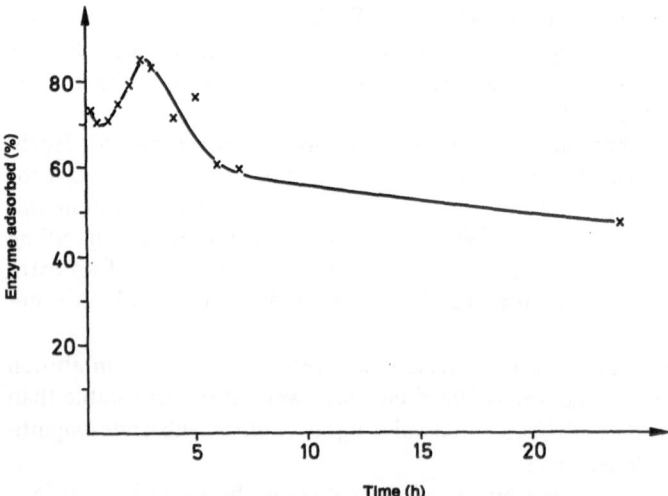

Fig. 5. Adsorption and desorption of cellulase to steam-exploded wheat straw during a 24 h hydrolysis. (Vallander and Eriksson [5]) Biotechnology and Bioengineering. Copyright (c) 1985, John Wiley & Sons, Inc. Reprinted by permission of John Wiley & Sons, Inc.

followed by an increase in adsorption, peaking at 3 h. This is followed by an equally rapid desorption of enzymes up to 6–7 h. Approximately 50% of the enzyme activity is found in the hydrolysate after 24 h.

Klyosov [43] pointed out that the ability of a cellulase to solubilize cellulose is directly related to its ability to adsorb to it. It has recently been shown that the adsorption capacities of a given type of cellulase from various microorganisms may differ by factors of as much as 100–1000 [43]. Considerable differences in adsorbability between isoenzymes from the same fungal strain are also usually observed. Klyosov concludes that "if a cellulase binds weakly to a crystalline cellulose this substrate is virtually resistant to hydrolysis by the enzyme in question, irrespective of the amount of enzyme used".

Stutzenberger [44] studied the adsorption of *endo*-glucanase on protein-extracted luzern fibres. Two of the three *endo*-glucanases from *Thermomonospora curvata* were rapidly adsorbed on the fibers while the third was not adsorbed at all. Adsorption was complete within 6 minutes. The rapid adsorption is in good agrement with other reports [5, 45].

Beldman et al. [46] also found that both the six *endo*- and the two *exo*-glucanases of a *T. viride* strain could be dividied into two groups adsorbing strongly and weakly, respectively, to crystalline cellulose. They concluded that the differences in adsorption will have implications for the recovery of the enzymes. These results raise the question of whether the appearance of both high- and low-adsorbing *endo*-glucanases within a strain is a property common to all cellulolytic organisms.

There is evidence that enzymes are adsorbed not only to the cellulosic part of the substrate. Adsorption to the lignin also takes place. Work by Sutcliffe and Saddler [47], Chernoglazov et al. [48] and Deshpande and Eriksson [49] show that cellulolytic enzymes bind strongly to lignin. Sutcliffe and Saddler [47] also observed that β-glucosidase, which does not bind to the polysaccharides, has a high affinity for various lignin fractions.

Chernoglazov et al. [48] studied the distribution of enzymes between various substrates and water. Cellulose and lignin from birch, larch and spruce wood were the substrates of choice. The binding strength was low to xylan, but high to both cellulose and lignin. A maximum of more then 90% adsorption of the *endo*-glucanase was obtained on the two latter substrates. Isolated lignin appears to inactivate the enzyme whereas this was not observed with the ligincarbohydrate complex. Inactivation was however observed when cellulases were adsorbed on steam-exploded lignocellulosic materials. This is probably due to the formation of inhibitors during steam explosion.

The question of reversibility and competition in adsorption of enzymes has only recently been addressed. Kyriacou et al. [50] studied the adsorption of a cellobiohydrolase (CBHI) and three *endo*-glucanases (EGI, EGII and EGIII) from *T. reesei*. Using [^{3}H)] and [^{13}C]-labeled enzymes, the adsorption and desorption of enzymes could be followed. For each of the enzymes studied using two types of label they showed that there is an exchange between adsorbed cellulase and that present in the bulk solution. If, however, liquid was filtered off from an enzyme-saturated cellulose suspension and a fresh, enzyme-free solution was added to the cellulose, no release of labeled protein into the bulk solution was detected. The CBHI fraction was preferentially adsorbed in competition with the *endo*-glucanases (EGI, II and III). EGI was also strongly adsorbed in combination with EGII and EGIII. It has been reported by others that the *T. reesei* cellobiohydrolases have the strongest adsorption affinity [51].

It has also been shown by Kyriacou et al. [50] that two enzymes added together are adsorbed to a greater extent than single enzymes. If two enzymes are fed in succession, the results show that once CBHI has become adsorbed it is not displaced by any other enzyme, but a great proportion of the other enzyme species is adsorbed as well. Both EGI and, particularly, CBHI are able to release already adsorbed EGII and EGIII. They conclude that the cellulose surface contains both distinct and common adsorption sites for the enzymes studied and that the extent to which they are occupied depends on the enzyme components involved and on the order in which they are added.

Hydrolysis was slightly impaired in the presence of NaCl (0.1 M), probably due to a reduced synergism between the cellobiohydrolase and the endoglucanase components [50].

The reason why microorganisms produce both strongly and weakly binding enzymes could be to facilitate hydrolysis of both the solid substrate and the solubilized oligosaccharides.

3.5 Enzyme Recirculation

Interest in the recovery of cellulolytic enzymes has been devoted mainly to enzymes in solution [4, 5, 52–56]. This may be adequate in the ideal case of a delignified substrate undergoing complete or nearly complete hydrolysis. If, however, lignocellulosic material is hydrolysed, the adsorbability of all enzyme species and their hydrolysing capacity as well as the presence of a residue after hydrolysis are factors that have to be considered in enzyme recovery.

It has been mentioned earlier in this report that the cellulases have a high affinity not only for cellulose but also for lignin [47–49]. Even if the polysaccharides of the substrate are completely hydrolysed, a significant portion of the enzymes i.e. those bound to the lignin, will be lost, at the end of the hydrolysis with the removal of the solid residue.

If only enzymes in solution are recovered, this fraction may consist of weakly binding enzymes with a limited capacity to hydrolyse a solid substrate. This enzyme mixture would consequently be of inferior quality compared to the starting mixture. This appears to be the case in a study reported by Mes-Hartree et al. [54] (Table 3). They studied enzyme recycling in a combined enzymatic hydrolysis and fermentation process. The assayed enzyme activity in the culture filtrate after CHF was less than 20 % of the original filter paper activity. The combination of 60 % of the recovered enzymes with 40 % fresh enzymes resulted in the production of 30 % less sugar than in the case where 100 % fresh enzymes were used.

Different approaches have been suggested to minimize the enzyme loss in a hydrolysis process. Extraction of the lignin from a steam-exploded substrate

Table 3. Hydrolytic activity of various proportions of recovered and fresh enzymes[a]

Original/recovered filtrate, %	Enzyme profile[b]			Hydrolysis products	
	Endoglucanase	β-Glucosidase	Filter paper	Reducing Sugars, mg ml^{-1}	Glucose, mg ml^{-1}
0/100	30.0	6.0	2.0	5.7	3.1
20/80	94.8	7.9	4.6	16.0	11.3
40/60	212.0	13.0	9.6	21.0	15.6
60/40	299.2	15.6	13.0	25.2	19.3
80/20	385.6	18.0	16.5	27.9	22.1
100/0	472.0	20.6	20.0	30.8	25.4

[a] Original culture filtrate had enzyme activity of 23.6 IU endoglucanase, 1.03 IU β-glucosidase, and 1.00 U filter paper activity. Hydrolysis was performed at 5 % cellulose concentration for 48 h at 45 °C.
[b] Enzyme profile given as Ug^{-1} substrate
(Mes-Hartree, Hogan and Saddler [54]) Biotechnology and Bioengineering. Copyright (c) 1987, John Wiley & Sons, Inc. Reprinted by permission of John Wiley & Sons, Inc.

with water and alkali prior to hydrolysis was unsuccessfully tried to solve the problem of the lignin-adsorbed cellulases [54]. Since the adsorbed cellulases have a superior hydrolysing capacity, it follows that the enzymes bound to the solid residue at the end of the hydrolysis ought to be recovered to the greatest possible extent. By washing the residue with phosphate buffer [42] or adjusting the pH to neutrality [55], some of the enzymes can be released. A drawback of such methods is the cost incurred by the purchase of the necessary chemicals and the dilution of the enzymes. Rao et al. [4] were able to recover more than 90 % of the Filter Paper Activity by grinding an enzym-treated bagasse residue with glass powder and extracting it with Tween-80. It is thus possible to desorb enzymes from a residue by various means but technically and economically suitable methods have still to be found.

Vallander and Eriksson [42] have shown that the bound enzymes can be recovered simply by bringing the residue into contact with fresh substrate, whereby a continued saccharification is achieved.

3.6 The Hydrolysis Process

The time scale of an enzymic hydrolysis process usually falls within the range of 24 to 72 hours. It appears that 48 hours or more are often necessary to achieve complete saccharification of the substrate. The rate of hydrolysis is high during the first few hours after the start and then rapidly declines. The decrease depends on several factors, competitive inhibition of the enzymes, depletion of substrate and its decreasing susceptibility to cellulolytic attack being the three most important. There are, however, different ways by which the reaction velocity can be influenced. A high enzyme-to-substrate ratio, for example, improves the rate of saccharification. For a given amount of enzyme and substrate this can be achieved by successive additions of substrate instead of adding it all at once.

Product inhibition is mainly caused by the presence of cellobiose and glucose. Addition of β-glucosidase to relieve the inhibition of the *endo-* and *exo-*glucanases by cellobiose has already been mentioned. The inhibitory effect of glucose can be minimized by the choice of combined hydrolysis and fermentation as the process design, but inhibition is apparently caused not only by sugars in solution but also, more particularly, by sugars adsorbed onto the substrate. This adsorbed sugar seems to create an inhibitory microenvironment around the cellulolytic enzymes even when the sugar concentration in the hydrolysate is low.

Vallander and Eriksson [5, 42] have shown that removal of the hydrolysate at an early or intermediate stage of hydrolysis and addition of fresh buffer to the substrate results in a greater saccharification than if the hydrolysis is allowed to run uninterrupted for the same length of time. The explanation of this effect seems to be that addition of fresh liquid washes sugar away from the substrate and thereby reduces competitive inhibition. This conclusion is supported by the observation that a significant quantitity of sugar, up to 10% of the weight of the substrate residue, can be released from the residue by a simple water wash (Fig. 6).

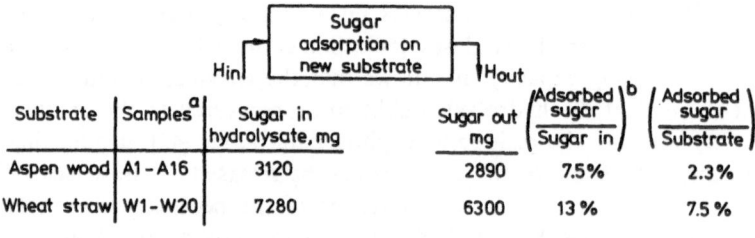

Substrate	Samples[a]	Sugar in hydrolysate, mg	Sugar out mg	$\left(\dfrac{\text{Adsorbed sugar}}{\text{Sugar in}}\right)$[b]	$\left(\dfrac{\text{Adsorbed sugar}}{\text{Substrate}}\right)$
Aspen wood	A1 – A16	3120	2890	7.5%	2.3%
Wheat straw	W1 – W20	7280	6300	13%	7.5%

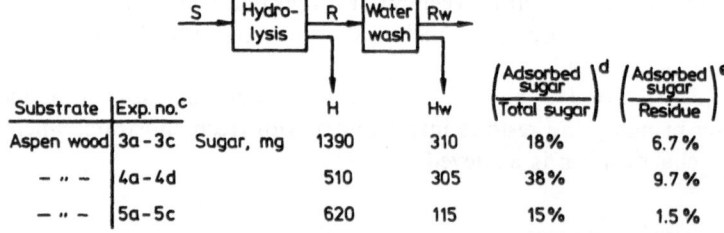

Substrate	Exp. no.[c]		H	Hw	$\left(\dfrac{\text{Adsorbed sugar}}{\text{Total sugar}}\right)$[d]	$\left(\dfrac{\text{Adsorbed sugar}}{\text{Residue}}\right)$[e]
Aspen wood	3a – 3c	Sugar, mg	1390	310	18%	6.7%
– " –	4a – 4d		510	305	38%	9.7%
– " –	5a – 5c		620	115	15%	1.5%

Fig. 6. The enzymes in the hydrolysate are recovered in contact with the new substrate. The figure shows the amount of sugar simultaneously adsorbed. After hydrolysis, part of the sugar is bound to the solid residue but it can be released by washing with water. (Vallander and Eriksson [42]) Enzyme Microb. Technol., 9 (1987), 714—720. Courtesy by BUTTERWORTH PUBLISHERS.

The positive effect of withdrawing hydrolysate is independent of the extent of hydrolysis. In a 48-h hydrolysis experiment with aspen wood and wheat straw, the saccharification yields increased from 53 and 49% to 67 and 56%, respectively, when the hydrolysate was removed after 24 h (Fig. 7) [42]. Removal of the hydrolysate after 15 min in a 24 h hydrolysis also resulted in a 9% improvement in the yield of saccharification compared with uninterrupted hydrolysis [5].

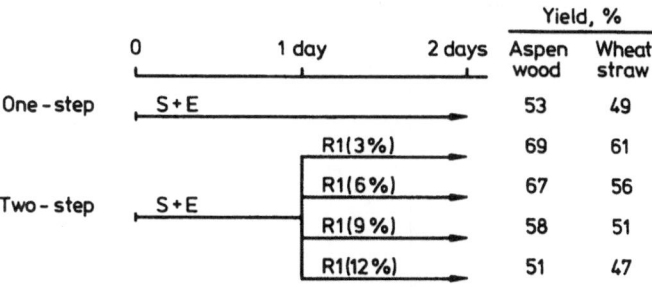

| | 0 | 1 day | 2 days | Yield, % | |
				Aspen wood	Wheat straw
One-step	S+E		▶	53	49
		R1(3%)	▶	69	61
Two-step	S+E	R1(6%)	▶	67	56
		R1(9%)	▶	58	51
		R1(12%)	▶	51	47

S = Substrate, E = Enzyme, R = Residue

Fig. 7. The saccharification yields are given as a percentage of the theoretical values. Hydrolyses were run at a substrate concentration of 6% in the first step ("$S + E$"). (Vallander and Eriksson [42]) Enzyme Microb. Technol., 9 (1987), 714—720. Courtesy by BUTTERWORTH PUBLISHERS

To investigate the potential advantages of hydrolysate removal and recirculation of enzymes bound to the residue and enzymes in solution, Vallander and Eriksson [16] studied two process models where the hydrolysate was collected after 2, 6 and 24 hours. The soluble and adsorbed enzymes were recirculated at the beginning of the process, where new substrate and a limited amount of fresh enzyme was added on days 2–4. As a reference model, a repeated 24-h continuous hydrolysis process was chosen. Enzymes in solution and those obtained from the solid residue by a simple water wash were recirculated to new substrate. The same amount of fresh enzymes was added daily to the reference as well to the previous process model systems. The reference system was clearly less efficient in hydrolysing the substrate during 4 consecutive days.

The rate of hydrolysis may also be increased by the use of membrane technology. Continuous ultrafiltration of the hydrolysate provides a way of removing the inhibiting glucose from enzymes in solution. High degrees of saccharification have been obtained in this way [39]. Reverse osmosis may also be a way of producing a sufficiently concentrated sugar solution instead of performing the hydrolysis at a high substrate concentration.

During the course of hydrolysis, part of the cellulases and of the β-glucosidases become desorbed due to solubilization of the substrate. These enzymes contribute to the conversion of water-soluble cellodextrins to glucose but at sufficiently high oligosaccharide concentrations *trans*-glycosylation reactions also occur. Ajisaka et al. [57] have shown that four kinds of glucose-disaccharides were formed in a 90% glucose solution when β-glucosidase was added. A study by Prakash et al. [58] reports the formation of oligosaccharides by *Trichoderma harzianum* incubated in a lactose medium. More than 30% of the added lactose was converted to trisaccharides and higher oligomers, the main product being 6'-galactosyllactose. Similar results have been reported by Mozaffar et al. [59] with *Bacillus circulans*.

Significant *trans*-glycosylation reactions have also been reported by Gusakov et al. [60]. At an initial cellobiose concentration of 0.1–0.3 M (approximately

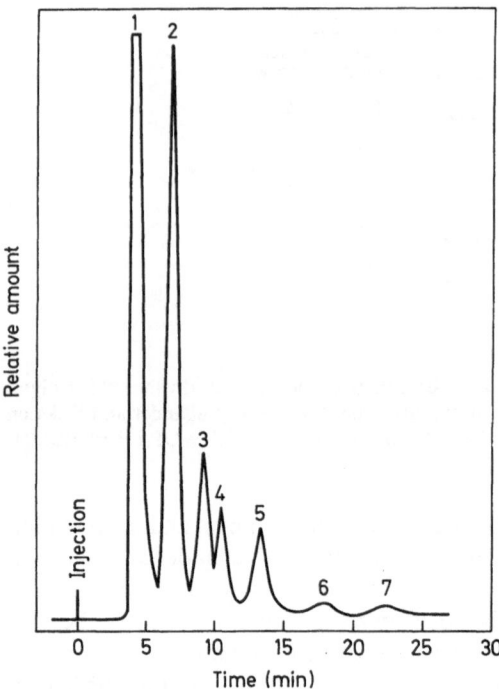

Fig. 8. HPLC analysis of cellobiose hydrolysis products. *1.* water; *2.* D-glucose; *3.* cellobiose; *4.* gentiobiose; *5.* isocellotriose; *6* and *7.* non-identified products. (Gusakov et al. [60]) Enzyme Microb. Technol., 6 (1984) 275–282. Courtesy by BUTTERWORTH PUBLISHERS.

$30\text{–}100 \text{ g l}^{-1}$) *trans*-glycosylation products amounted to 25–30% of the total saccharide content, (Fig. 8). Already at cellobiose concentrations of 0.01 M and higher, *trans*-glycosylation products were formed.

Streamer et al. [61], also showed that 3 of the 5 *endo*-glucanases present in culture solutions of *Sporotrichum pulverulentum* produced cellotriose when incubated with cellobiose (2% conc.), and cellotetraose + cellopentaose when incubated with cellotriose (2% conc.). *Trans*-glycosylation reactions are a function of substrate concentration and, on the surface of a solid substrate where microenvironments with a high sugar concentration are created, both competitive inhibition and *trans*-glycolysation reactions are with certainty a hindrance to fast hydrolysis.

This being the case, it follows that enzymic hydrolysis can be improved by adjusting the enzyme mixture composition during the course of hydrolysis in order to reduce unproductive side-reactions.

To conclude, different process models or steps have been reviewed which deal with the problem of increasing the rate of hydrolysis. It might be fruitful to combine some of these models and techniques to construct a process model which would be more efficient than any one of them alone. Such a process model could include a hydrolysis step supplemented with β-glucosidase, and multiple removal of the hydrolysate combined with its passage through an ultrafiltration unit to remove the sugars produced and to recirculate soluble enzymes. Enzyme recovery should also include recirculation of a major part of the solid residue.

3.7 Stability of Cellulolytic Enzymes

Several factors influence the stability of enzyme. They can be inactivated by too high temperatures, by experimental conditions in the hydrolysis vessel (other than temperature) and by the presence of inhibitors. This does not include the competitive inhibitors formed during hydrolysis, which have been discussed earlier in this review. In contrast to these destabilizing factors, the immobilization of enzymes is a technique for improving enzyme stability.

3.7.1. Temperature

Data on the thermal stability of *endo*-glucanases, cellobiohydrolases and β-glucosidases in terms of their half-life, $t_{0.5}$ (i.e. the length of time to reduce the enzyme activity by 50%) are presented in Table 4. It is obvious that the stability of a given type of enzyme varies considerably for enzymes produced by different strains. Inactivation has been studied at temperatures between 45 and 55 °C.

endo-Glucanases appear to be most stable in the sense that none of them has a half-life below 40 h at 50 °C (Table 4). In general, their half-lives exceed 2 days. The average value is practically the same as for the cellobiohydrolases, which is 90 h.

The cellobiohydrolases (Table 4) have half-lives ranging from 11 to 289 h, with an average value of 85 h. It appears remarkable that several of these enzymes are fairly unstable, 4 out of 11 having a $t_{0.5}$ of less than 30 h, although, in the presence of substrate the stability of the enzymes is usually significantly improved.

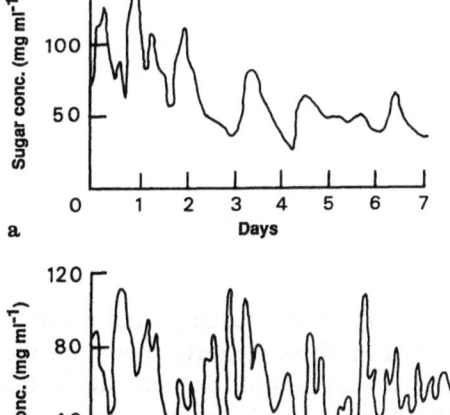

a

b

Fig. 9a, b. The long term effect of temperature on hydrolysis rates. **a** Continuous cellulose hydrolysis was carried out at 42 °C for one week. **b** Cellulose hydrolysis was carried out at 37 °C for two weeks under identical conditions except for the temperature. (Tan et al. [33]) Appl. Microbiol. Biotechnol, 26 (1987), 21—27. Courtesy by Springer-Verlag.

Table 4. Thermal stability of cellobiohydrolases (CBH), *endo*-glucanases (EG) and β-glucosidases (β-GL) in the temperature range of 45 to 55 °C

Enzyme		Half-life	Temperature	Remark	Reference
Type	Source[a]	h	°C		
CBH	T.v. (BDH)	122	45		Cantarella
EG		133	45		et al. [62]
β-GL		46	45		
CBH	T.v. (Miles)	29	45		
EG		88	45		
β-GL		61	45		
CBH	T.r.	289	45		
EG	(Celluclast)	41	45		
β-GL		—	45		
CBH	Cellulase	11	45		
EG	AP	577	45		
β-GL		198	45		
CBH	Cellulase	21	45		
EG	AP3	60	45		
β-GL		103	45		
CBH	A. niger	18	45		
EG		91	45		
β-GL		117	45		
β-GL		179	50	Reported as	Dekker [63]
β-GL		55	50	>48 h, by Dekker	
CBH	C30	186	50	The half-life is	Reese and
	NG14	131	50	the mean value of	Mandels [53]
	6a	88	50	two estimates based	
	9414	76	50	on logarithmic in-	
	MCG77	85	50	activation	
β-GL		10	45		Gianfreda
		5	50		et al. [64]
CBH		85			
EG	Median				
β-GL	values	90			
		82			

[a] T.v. = *Trichoderma viride*

The range of half-lives of β-glucosidases is also wide, ranging from 5–198 h, the average being 82 h. The results presented in Table 4 clearly show that the stability of some β-glucosidases drastically decreases with temperature in the vicinity of 50 °C. This is also true for the *T. viride endo*-glucanase studied by David and Thiry [65].

It can thus be concluded that cellulase stability under normal hydrolysis conditions, i.e. pH 5 and 50 °C, is satisfactory as long as enzymes from a proper strain are chosen.

Another strategy to maintain long-term activity is to reduce the hydrolysis temperature from 50 °C by some 10 degrees. It has been shown by Tan et al. (Fig. 9) [33], that a constant level of saccharification could be maintained for two weeks at 38 °C but for only one week at 42 °C. By lowering the hydrolysis temperature, the enzyme stability could thus be increased and, if factors other than the enzyme itself are rate-limiting, the long-term rate of hydrolysis would not be adversely affected.

3.7.2. Other Experimental Conditions

Apart from temperature, experimental variables such as pH, concentration of substrate, agitation [53, 66], shear forces [66], air in the gas phase [66] and ionic strength [67] are important for the activity and stability of enzymes.

Kim et al. [66] studied inactivation of cellulases exposed to an air-liquid interface and shear forces. When subjected to both factors, inactivation was more severe than in the case of shear forces alone. Without agitation, the *endo*-glucanase activity decreased only 10% in 5 days. The inactivation was considered to be due to unfolding of the enzyme-protein at the air-liquid interface. When a surfactant was added, the inactivation decreased because of limited access of enzyme molecules to the surface.

Reese and Mandels [53] observed a decrease in hydrolysis yield with *T. reesei* C-30 cellulase due to shaking. They also found that the enzyme activity in solution after hydrolysis was somewhat higher if the gas atmosphere consisted of CO_2 instead of air (unshaken conditions).

Castanon and Wilke [68] found that hydrolysis and enzyme recovery were improved in the presence of the surfactant Tween 80. They ascribed the effect to a reduction in enzyme adsorption onto the substrate. A higher enzyme activity was thus observed in solution in the presence of Tween 80.

It appears unlikely, however, that a reduction in the amount of adsorbed enzymes would increase saccharification. A more plausible explanation is provided by Kim et al. [66], i.e. that the surfactant reduces the inactivation of enzymes in solution.

It should also be observed that Rao et al. [4] in their experiment to recover cellulases from bagasse residue, obtained better results in the presence of Tween 80. Obviously there are several factors affecting enzyme stability and activity negatively. It is pertinent to ask which of them are the most important ones. Gusakov et al. [69] concluded from a computer simulation of cellulose hydrolysis in a batch-stirred reactor that thermal inactivation of the cellobiose-producing enzymes and inhibition by cellobiose are the main factors affecting the hydrolysis kinetics negatively. We agree to this but would like to add that competitive inhibition by glucose is also important.

3.7.3. Immobilization

Immobilization of β-glucosidases has been suggested as a way to achieve two objectives. First, to improve enzyme stability and, second, to facilitate a decrease in enzyme consumption.

Immobilization of β-glucosidase from *A. phoenicis* was studied by Sundstrom and coworkers [70]. When alumina of controlled pore size was used as the solid support, 90% of the enzyme activity was retained on immobilization. The activity was well preserved, only 10% being lost over a period of 500 h, although the thermal stability of the immobilized enzyme was no better than that of the soluble enzyme. A β-glucosidase isolated from sweet almonds was immobilized by Venardos et al. [71] on Amberlite DP-1, a cation-exchanger. The Michaelis-Menten constant, K_m, increased with increasing enzyme loading as a result of competitive inhibition by the enzyme itself due to blocking of some of its active sites.

The stability of the immobilized enzyme preparation was good, half-lives being found in the range of 200–375 h. On the other hand, less than 50% of the soluble enzyme activity at pH 5 was retained upon immobilization.

Aspergillus niger β-glucosidase was immobilized on three types of support, namely Concanavalin A-Sepharose, CNBr-Sepharose and microcrystalline cellulose [72]. The enzyme-support complexes retained 80–90% of the bound activity. The pH-optimum was not changed by immobilization and the pH stability was not improved with the exception of that of the CNBr-Sepharose-enzyme complex. An improved thermal stability was obtained at 65 °C when CNBr-Sepharose- and Concanavalin A-Sepharose-enzyme complexes were treated with glutaraldehyde. Enzyme recovery was fairly good. After several experiments with the same CNBr-Sepharose enzyme preparation, 54% of the activity was left, despite losses in handling and washing of the enzyme.

To conclude, β-glucosidase can be efficiently immobilized on various types of support, retaining a high degree of activity. The half-lives of these preparations are usually several hundred hours. The thermal stability varies. It may be the same as that of soluble β-glucosidase or better. However, these advantages must be balanced against an often decreased reaction rate and increased sensitivity to competitive inhibition of the immobilized enzymes.

3.7.4. Inhibitors

Cellulases are remarkably stable to changes in pH and temperature and to chemical inhibitors [73]. Inhibitors may, however, be formed during pretreatment of the raw material. In the case of steam pretreatment, it is known that part of the polysaccharides are degreaded to furfural and hydroxymethyl furfural, which are known inhibitors of the cellulases. Mes-Hartree and Saddler [74] studied the inhibition of the cellulases by these compounds. They were found not to be inhibitory at concentrations normally found in steam-exploded lignocellulosic materials, although tannins and substituted phenols such as chlorophenol and others, are known to inhibit the action of the cellulases [73]. It has also been shown [75] that the β-glucosidase of the rumen bacterium *Bacteroides succinogenes* is inhibited by vanillin, by ferulic acid and particularly by *p*-coumaric acid.

Reese and Mandels [53] have shown that the cellobiohydrolase but not the *endo*-glucanase from *T. reesei* is very sensitive to inhibition by merthiolate. It was shown by Mes-Hartree and Saddler [74] that the β-glucosidase activity is hampered by inhibitory substances present in steam exploded material. It is however sufficient to wash the steam-pretreated material prior to enzymic hydrolysis to avoid this inhibition. Since the wash water also contains monomeric and oligomeric xylose, one has to consider the possible inhibition of the xylose-bioconverting organism if this water is to be bioconverted.

Inhibitors may be formed not only during steam pretreatment of lignocellulosic materials. Dekker [76] reports that a specific β-glucosidase inhibitor is formed during the enzymic hydrolysis of *Eucalyptus regnans* pretreated by autohydrolysis. The inhibitor was active against the β-glucosidase of *T. reesei* C-30, but the corresponding enzyme from *A. niger* was unaffected. It was also shown that the Filter Paper Activity and *endo*-glucanase activity of *T. reesei* C-30 were not affected by this inhibitor.

Pfeifer et al. [77] studied the inhibition of *Saccharomyces carlsbergensis* W34 by some low-molecular degradation products from lignocellulosic materials. They found that hydroxymethyl furfural, furfural, methylglyoxal and lignin-related compounds like phenol and vanillin significantly inhibited the bioconversion in the concentration range from 1 to 10 mg ml^{-1}. It thus appears that inhibitors formed in the pretreatment mostly affect the bioconversion step and that the major inhibition problem during enzymic hydrolysis is competitive inhibition by cellobiose and glucose.

4 Part C: Bioconversion ("Fermentation")

In the bioconversion of sugars obtained by pretreatment and enzymic hydrolysis of lignocellulosic materials, most of the problems are related to the bioconversion of xylose, since bioconversion of glucose to ethanol is a well established technology.

Topics of interest include:
— which sugars can be fermented to ethanol,
— yield,
— productivity,
— ethanol and sugar tolerance,
— inhibitors (other than ethanol and sugar),
— process conditions.

It has been shown by du Preez et al. [78] that *Candida shehatae* and *Pichia stipitis* are able to convert glucose, mannose, galactose and xylose to ethanol but not xylitol, arabinose or rhamnose. In contrast to *C. shehatae*, *P. stipitis* is also able to convert cellobiose. *P. stipitis* is the more efficient of the two strains, giving a higher ethanol yield (0.39 g g^{-1} xylose) and higher productivity.

Xylose bioconversion has recently been reviewed by Skoog and Hahn-Hägerdal [79]. Only a short summary of the present state of the art will therefore be given here.

Technically, ethanol can be produced from xylose in two ways. Xylose is either directly converted to ethanol or is first converted to xylulose by xylose isomerase and then to ethanol. The review [79] presents results from 6 studies screening a

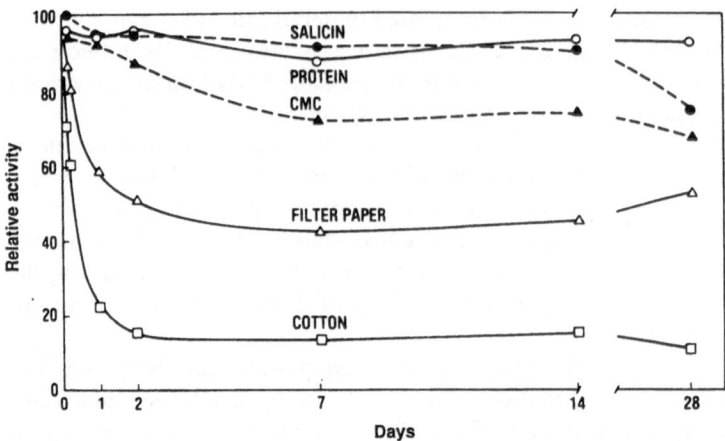

Fig. 10. Stability of cellulase components of *T. reesei* QM 9414. The enzyme was incubated at pH 4.8 and at 50 °C (in the presence of 0.01 % merthiolate) (●– – –●) β-Glucosidase; (△– – –△) endo-β-glucanase; (△——△; ——) combined action of endo-glucanase and cellobiohydrolase. (Reese and Mandels [53]) Biotechnology and Bioengineering. Copyright (c) 1980. John Wiley & Sons, Inc. Reprinted by permission of John Wiley & Sons, Inc.

total of 87 yeast strains for xylose bioconverting ability. Several strains of *Brettanomyces*, *Candida* and *Pichia* were included in these studies. The xylulose bioconverting organisms belonged mainly to the genera *Candida*, *Saccharomyces* and *Schizosaccharomyces*.

The ethanol concentrations obtained varied considerably between strains as did the formation of the by-product xylitol. The range of ethanol concentrations observed with xylose- and xylulose-converting organisms were however practically the same. Concentrations of 2–3 % were obtained with the most efficient strains, but xylitol formation of up to 2 % was also encountered.

A limitation of the xylose isomerase method is that the enzyme is restricted by an equilibrium reaction, by which no more than 20 % of the xylose is transformed to xylulose. This problem may, however, be overcome by using the concept of simultaneous isomerization and bioconversion [79]. At a cell density of 75 g l^{-1} (dw) of baker's yeast, 62 g l^{-1} of ethanol was obtained in a fed-batch culture. The yield was 0.34 g ethanol per g of xylose consumed and the productivity amounted to 1.25 g l^{-1} h^{-1}. These are encouraging results even in comparison with what is normally obtained in the bioconversion of hexoses, viz. 50 g l^{-1}, 0.5 g g^{-1} and 2 g l^{-1} h^{-1}, respectively [79].

It is not as simple to bioconvert hydrolysates from lignocellulosic materials. Skoog and Hahn-Hägerdal [79] write: "Several studies report on product concentrations in the range of 20–30 g l^{-1}, yields higher than 0.4 g g^{-1} and productivities higher than 0.5 g l^{-1} h^{-1}". Obviously, efforts to increase ethanol concentration and productivity when bioconverting lignocellulosic hydrolysates are welcome.

Tolan and Finn [80] studied the bioconversion of xylose and arabinose by *Erwinia* species. These are gram-negative facultative anaerobic bacteria. It is worth mentioning that they were able to bioconvert arabinose, which is normally present in lignocellulosic substrates. *E. chrysanthemi* B374 was particularly interesting because of its high ethanol tolerance, i.e. 4%. When the pyruvate decarboxylase (pdc) gene from *Zymomonas mobilis* was introduced into *E. chrysantemi* B374, the ethanol yield increased from 0.7 up to 1.45 mol mol^{-1} xylose. A decrease in the formation of formate, acetate and lactate was also observed, but these advantages were obtained at the cost of a reduced ethanol tolerance i.e. 2%.

Similar results were also reported by Tolan and Finn [81] in the bioconversion of xylose with *Klebsiella planticola*, which was also carrying the pdc gene from *Zymomonas mobilis*. An ethanol yield of 1.3 mol mol^{-1} xylose was achieved, the by-product formation was very small at pH 5.4 (increased with higher pH) and, positively, the ethanol tolerance was maintained at 4%.

According to Chung and Lee [82], the theoretical ethanol yield from xylose is 1.67 mol mol^{-1} xylose (or 9.51 g g^{-1}). The above mentioned values, obtained with the transformed *E. chrysanthemi* were 1.45 and 1.3 mol mol^{-1}, thus representing 87, and 78% respectively of the theoretical yields.

A few recent studies deal with the effect of oxygen on xylose bioconversion but other hydrogen acceptors have also been studied. The importance of oxygen on the growth and bioconversion of xylose differs for various yeast strains. Delgenes et al. [83] studied *P. tannophilus*, *P. stipitis*, *Kluyveromyces marxianus* and *Candida shehatae*. All four strains were able to metabolize xylose aerobically, but without the formation of ethanol. Only *P. tannophilus* and *P. stipitis* converted xylose to ethanol anaerobically, the latter strain at a yield of 0.4 g g^{-1} xylose. Under micro-aerophilic conditions *K. marxianus* also produced ethanol, although less efficiently than the two other strains. When all the xylose had been consumed and the ethanol concentration was at its peak, all three strains started to use ethanol for growth.

It is known that xylose bioconversion is sensitive to oxygen concentration. A problem which remains to be solved is how to measure and control the oxygen level effectively. This question has been addressed by du Preez et al. [84], who regulated the aeration by varying the stirrer speed using the redox potential as a control variable. The critical dissolved oxygen tension (DOT) for optimal xylose bioconversion coincided with the sensitivity limit of the DOT electrode. Xylose bioconversion by *C. shehatae* was most effective at a redox potential maintained at 300 ± 10 mV, but regulation of the oxygen tension by monitoring the redox potential proved ineffective in the case of *P. stipitis*.

Ligthelm et al. [85] have shown that bioconversion of xylose by *P. tannophilus* is influenced by the age of the culture used. Cultures in their late exponential and stationary phases are more succesful in fermentation than cultures harvested from the lag and early exponential phases. These authors were able to relate the differences in bioconversion efficiency to the concentrations of NADH- and NADPH-linked xylose reductase and to the levels of NAD- and NADP-linked xylitol dehydrogenase.

It is generally agreed that the positive effect of oxygen on xylose bioconversion is dependent upon the regeneration of cofactors for the conversion of xylose to

90 L. Vallander and K.-E. L. Eriksson

xylulose via xylitol. Accumulation of xylitol during bioconversion is due to a deficit of the NAD-cofactor necessary for the production of xylulose. Oxygen is thus needed to restore the NAD level by oxidation of NADH. Ligthelm et al. [86] have demonstrated that some hydrogen acceptors other than oxygen can also be used successfully for this purpose (Fig. 11). They reported that the molar yields of ethanol increased from an initial level of 1.03 to 1.63, 1.43 and 1.24 when acetaldehyde, acetone and acetoin, respectively were added to the culture. The yield of xylitol decreased simultaneously.

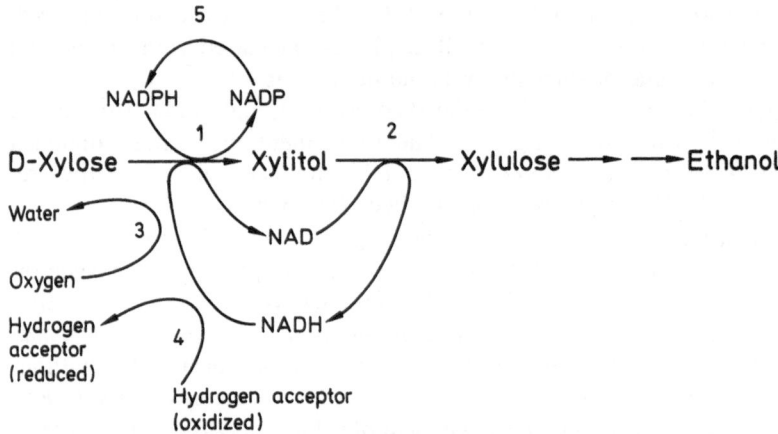

Fig. 11. Pathway of D-xylose utilization by yeasts. Enzymes: xylose reductase (*1*), xylitol dehydrogenase (*2*), mitochondrial electron transport system (*3*), dehydrogenase system (*4*) and pentose phosphate pathway (*5*). (Ligthelm et al. [86]) Biotechnology and Bioengineering, Copyright (c) 1989, John Wiley & Sons, Inc. Reprinted by permission of John Wiley & Sons, Inc.

In an experiment with oxygen as the hydrogen acceptor, the molar ethanol yield increased from 0.97 to 1.43 per mol of xylose consumed [86], but the level of xylitol was unaffected in this case. Ligthelm et al. also observed [86] that far more ethanol was produced per mol of oxygen consumed than one would expect from the stoichiometry of xylose bioconversion. This indicates that oxygen is not only used by *P. tannophilus* as a hydrogen acceptor but serves other unknown purposes as well. The role of oxygen during xylose bioconversion thus needs to be further elucidated.

High values of yield and ethanol tolerance have been reported by Wayman and Parekh [87], when *C. shehatae* ATCC 2298 was used under semiaerobic conditions to convert hydrolysates from whole barley, containing approximately 70% of glucose and 30% of xylose. An ethanol concentration of 100 g l^{-1} had been reached in the bioconversion of a 260 g l^{-1} sugar solution when the bioconversion ceased, but the bioconversion resumed when the ethanol was removed by vacuum distillation. Ethanol yields in the range of 92 to 98% of the theoretical values were reported. Bioconversion of a 180 g l^{-1} sugar solution yielded 84 g l^{-1}

of ethanol (91 % yield of the theoretical value) and the bioconversion was complete within 72 h [87]. This is equal to an ethanol tolerance of 8 % or more during the xylose bioconversion, since almost all glucose had been converted before the xylose started to be utilized.

The great differences in ethanol yield and ethanol tolerance in pentose bioconversion reported by different groups may have several explanations. The choice of substrate and its pretreatment, the presence of inhibitors, the sugar composition of pretreatment, the presence of inhibitors, the sugar composition of the hydrolysate, the choice of organism and the particular strain being used and finally the adaptation of the strain to the medium are some factors of possible importance.

When assessing the ethanol tolerance of a strain, the experimental conditions must be considered. Novak et al. [88], who studied the inhibitory effect of ethanol on the organisms used during alcohol production, demonstrated that ethanol produced during batch cultivation is more inhibitory than added ethanol. These differences in ethanol inhibition are best explained by differences in the intracellular ethanol concentration.

Ando and coworkers [89] identified aromatic monomers present in steam-exploded poplar wood and assessed their influence on the bioconversion of the hydrolysate by *Saccharomyces cerevisiae*. The concentration of the aromatic monomers was 0.1 g l^{-1} and the conversion time was 50 h. The aromatic compounds contained different functional groups, such as $-OH$, $-OCH_3$, $-CHO$ and $-COOH$. Each functional group was given a number indicating its contribution to the inhibitory effect of the substituted aromatic compound. When these numbers were applied by us to monomeric substituted phenols which had been obtained in an inhibition study by Nishikawa et al. [90], a fairly good agreement in ranking was obtained between the actual inhibitory effect and the postulated inhibitory effect, at an inhibitor concentration of 0.4 g l^{-1} and a conversion time of 24 h. In accordance with this, Tran and Chambers [91] found that the presence of methoxyl groups reduce the toxicity and that aromatic acids are less toxic than the corresponding aldehydes.

It is, however, necessary to be cautious when using the method of Ando et al. [89] to assess the inhibitory effect. The method takes into consideration neither the inhibitor concentration nor the effect of time. The organism may adapt itself to the inhibitor and degrade it and the rate of bioconversion will then again increase. It thus appears that this method is limited to give a first estimate of the inhibitory effect of substituted aromatic phenols.

The ethanol tolerance of an organism appears to be improved by the addition of lipids to the growth medium. The effect has been explained by the incorporation of the lipid into the cellular membrane which helps the organism to sustain higher levels of ethanol. By adding of a mixture of ergosterol, linoleic acid and Tween-80 to a culture of *P. tannophilus*, Dekker [92] increased the ethanol yield from 0.20 to 0.32 g g^{-1} xylose. Even though the ethanol yields were low, the results suggest a possible way of improving ethanol tolerance.

5 Suggestions for Future Research

The great interest in developing an enzymic hydrolysis process reflects the wish to create an efficient and environmentally more suitable process than the old acid hydrolysis. This may also be part of a more general concern for the environment and the energy situation and a desire to switch to the use of renewable materials for the production of fuels and other chemicals. Plans for future research in this field and the evaluation of results are therefore likely to be concerned not only with the efficiency of certain process steps but also with the environmental and energy aspects which are becoming more and more important.

5.1 Pretreatment

Even though steam pretreatment is the favoured pretreatment, other ideas should be further exploited. Chemical pretreatment often creates disposal problems, but what can be achieved with different oxygen species such as oxygen, ozone and hydrogen peroxide?

The final yield of pentoses and hexoses varies with temperature of the steam pretreatment. The temperature to obtain maximum yields is not the same for the two types of sugars. Can this be improved in a two-step process pretreatment?

Steam pretreatment produces inhibitors. How is the formation of these inhibitors related to the pretreatment conditions?

The energy efficiency of steam pretreatment (steam consumption) is not well known. What are the prospects for energy recovery?

Xylan degradation products produced by steam pretreatment are mainly in the form of xylo-oligomers. How can the yield of monomeric xylose be improved?

5.2 Enzymic Hydrolysis

We have presented some ideas for how to improve the enzymic hydrolysis process. Several components were included in the suggested model. Variables to be considered are for instance:

a) Single or multiple additions of substrate;
b) Enzyme-to-substrate ratio;
c) The points in the process where hydrolysates are removed;
d) How should the distribution of enzyme additions be made?
e) How much of the solid residue should be recirculated?

Continued studies of the use of membrane technology in combination with enzyme hydrolysis are of interest, e.g. ultrafiltration for the separation of sugars from enzyme and reverse osmosis for the concentration of dilute sugar solutions. This involves both technical and economical considerations. Which flux can be obtained in a process with pretreated lignocellulosic materials as substrate? Problems with fouling how can they be overcome?

Would it be worthwhile to produce a fairly dilute sugar solution and to concentrate it by reverse osmosis before bioconversion instead of producing the final sugar concentration in the hydrolysis step alone?

If a hydrolysis temperature lower than 50 °C were chosen the enzyme would be stable for a longer time but the rate of hydrolysis would also decrease. How would this influence the size of the bioreactor?

What can be done to the residue removed? How much adsorbed sugar and enzymes can be recovered by washing? What is the character of the lignin in the residue compared to that of the lignin directly after steam pretreatment?

5.3 Bioconversion

The following points require further study:
a) Improvement of pentose converting organisms in terms of yield, productivity and ethanol tolerance.
b) Adaptation of strains to inhibitors.
c) Conversion of xylose and glucose together or separately.
d) Elucidation of the role of oxygen particularly during pentose bioconversion.

6 References

1. Reczey K, Laszlo E and Hollo J (1986) Starch. 38, 306
2. Mes-Hartree M and Saddler JN (1983) Biotechnol. Lett., 5(8), 531
3. Wilke CR, Yang RD, Sciamanna AF and Freitas RP (1981) Biotechnol. Bioeng., 23, 163
4. Rao M, Seeta R and Deshpande V (1983) Biotechnol. Bioeng., 25, 1863
5. Vallander L and Eriksson KE (1985) Biotechnol. Bioeng., 27, 650
6. Saddler JN, Brownell HH, Clermont LP and Levitin N (1982) Biotechnol. Bioeng., 24, 1389
7. Bonn G, Hörmeyer HF, and Bobleter O (1987) Wood Sci. Technol. 21, 179
8. Dekker RHF and Wallis AFA (1983) Biotechnol. Bioeng., 25, 3027
9. Rolz C, de Arriola MC, Valladares J and de Cabrera S (1987) Process Biochemistry, 22, 17
10. Clark TA and Mackie KL (1987) J. Wood Tech., 7 (3), 373
11. Brownell HH, Yu EKC and Saddler JN (1986) Biotechnol. Bioeng., 28, 792
12. Brownell HH and Saddler JN (1987) Biotechnol. Bioeng., 29, 228
13. Mamers H and Menz DNJ (1984) Appita 37(8), 644
14. Morjanoff PJ and Gray PP (1987) Biotechnol. Bioeng., 29, 733
15. Puls J, Poutanen K, Körner HU and Viikari L (1985) Appl. Microbiol. Biotechnol., 22, 416
16. Vallander L and Eriksson KE (1989) Biotechnol. Bioeng., under review
17. Dale BE and Moreira MJ, Paper presented at the 4th Symposium on Biotechnology in Energy Production and Conservation, May 11–14, 1982, Gatlinburg, Tennessee
18. Grethlein HE (1985) Biotechnology, 3, 155
19. Grous WR, Converse AO and Grethlein HE (1986) Enzyme Microb. Technol., 8, 274
20. Wong KKY, Deverell KF, Mackie KL, Clark TA and Donaldson LA (1988) Biotechnol. Bioeng., 31, 447
21. Chen HC and Grethlein HE (1988) Biotechnol. Lett., 10(12), 913
22. Rabinovitch ML, Nguyen Van Viet, Klyosov AA and Freitas RP (1981) Biotechnol. Bioeng., 23, 163
23. Ford CW (1983) Aust. J. Agric. Res., 34, 241

24. Cunningham RL, Detroy RL, Bagby MO and Baker FL (1981) Transactions of the Illinois state academy of science, 74(3—4), 67
25. Lee SB, Shin HS, Ryu DDY and Mandels M (1982) Biotechnol. Bioeng., 24, 2137
26. Eriksson KE, Hollmark BH and Pettersson A (1969) Svensk Papperstidn. 72, 551
27. Eriksson KE and Wood TM (1985) Biodegradation of Cellulose. In "Biosynthesis and Bio-degradation of Wood Components" (T. Higuchi, Ed) pp 469–503. Academic Press, London
28. Knowles J, Teeri T, Lehtovaara P, Pentila M and Saloheimo M (1988) The use of gene tech-nology to investigate fungal cellulolytic enzymes. In "Biochemistry and Genetics of Cellulose Degradation" (Aubert JP, Beguin P and Millet J, Eds) pp 153–169. Academic Press, London
29. Teeri TT, Lehtovaara P, Kauppinen S and Salovuori I (1987) Gene 51, 43
30. Poutanen K (1988) Diss. Techn. Res. Centre, Finland. Publication 47
31. Dekker RFH (1985) Biodegradation of the hemicelluloses. In "Biosynthesis and Biodegrada-tion of Wood Components" (T. Higuchi, Ed) pp 505–533. Academic Press, London
32. Reese ET, Shibata Y (1965) Can J Microbiol 11, 167
33. Tan LUL, Yu EKC, Mayers P and Saddler JN (1987) Appl. Microbiol., 26, 21
34. Dekker RFH, Karageorge H and Wallis AFA (1987) Biocatalysis, 1, 47
35. Perez J, Wilke CR and Blanch HW. Paper presented at the second chemical congress of the north American continent, Las Vegas, Nevada (1982)
36. Esterbauer H, Hayn M, Jungschaffer G, Taufratzhofer E and Schurz J (1983) J. Wood Chem. Technology, 3(3), 261
37. Ghosh P, Pamment NB and Martin WRB (1982) Enzyme Microb. Technol., 4, 425
38. Sternberg D, Vijayakumar P and Reese ET (1977) Can. J. Microbiol. 23, 139
39. Herr D (1980) Biotechnol. Bioeng., 22, 1601
40. Khan AW, Chin A and Baird S (1985) Biotechnol. Lett., 7(6), 447
41. Fadda MB, Dessi MR, Maurici R, Rinaldi A and Satta G (1984) Appl. Microbiol. Biotechnol., 19, 306
42. Vallander L and Eriksson KE (1987) Enzyme Microb. Technol., 9, 714
43. Klyosov AA (1986) Appl. Biochem. Biotechnol. 12, 249
44. Stutzenberger F (1987) Lett. Appl. Microbiol. 5, 1
45. Ghose TK and Bisaria VS (1979) Biotechnol. Bioeng., 21, 131
46. Beldman G, Voragen AGJ, Rombouts FM, Searle-van Leeuwen MF and Pilnik W (1987) Biotechnol. Bioeng., 30, 251
47. Sutcliffe R and Saddler JN (1986) Biotechnol. Bioeng. Symp., 17, 749
48. Chernoglazov VM, Ermolova OV and Klyosov AA (1988) Enzyme Microb. Technol., 10, 503
49. Deshpande MV and Eriksson KE (1984) Enzyme Microb. Technol., 6, 338
50. Kyriacou A, Neufeld RJ and MacKenzie CR (1989) Biotechnol. Bioeng., 33, 631
51. Ryu DDY, Kim C and Mandels M (1984) Biotechnol. Bioeng., 26, 488
52. Castanon M and Wilke CR (1980) Biotechnol. Bioeng., 22, 1037
53. Reese ET and Mandels M (1980) Biotechnol. Bioeng., 22, 323
54. Mes-Hartree M, Hogan CM and Saddler JN (1987) Biotechnol. Bioeng., 30, 558
55. Clesceri LS, Sinitsyn AP, Saunders AM and Bungay HR (1985) Appl. Biochem. Biotechn., 11, 433
56. Toyama N, Ogawa K and Toyama H (1983) Bull Fac. Agr. Miyazaki Uni., 30, 57
57. Ajisaka K, Nishida H and Fujimoto H (1987) Biotechnol. Lett. 9(4), 243
58. Prakash S, Suyama K, Itoh T and Adachi S (1987) Biotechnol. Lett., 9(4), 249
59. Mozaffar Z, Nakanishi K and Matsuno R (1988) Biotechnol. Lett., 10(11), 805
60. Gusakov AV, Sinitsyn AP, Goldsteins GH and Klyosov AA (1984) Enzyme Microb. Technol., 6, 275
61. Streamer M, Eriksson KE and Pettersson B (1975) Eur. J. Biochem., 59, 607
62. Cantarella M, Gallifuoco A, Scardi V and Alfani F (1984) Annals N.Y. Acad. Sci., 434, 39
63. Dekker RFH (1986) Biotechnol. Bioeng., 28, 1438
64. Gianfreda L, Livolsi AM, Scarfi MR and Greco Jr G (1982) Enzyme Microb. Technol., 4, 322
65. David C and Thiry P (1981) Eur. Polym. J., 17, 957
66. Kim MH, Lee SB, Ryu DDY and Reese ET (1982) Enzyme Microb. Technol., 4, 99

67. Dale BE and White DH (1983) Enzyme Microb. Technol., 5, 227
68. Castanon M and Wilke CR (1981) Biotechnol. Bioeng., 23, 1365
69. Gusakov AV, Sinitsyn AP and Klyosov AA (1987) Biotechnol. Bioeng., 29, 906
70. Sundstrom DW, Klei HE, Coughlin RW, Biederman GJ and Brouwer CA (1981) Biotechnol. Bioeng., 23, 473
71. Venardos D, Klei HE and Sundstrom DW (1980) Enzyme Microb. Technol., 2, 112
72. Woodward J and Wohlpart DL (1982) J. Chem. Tech. Biotechnol., 32, 547
73. Mandels M and Reese ET (1965) Ann. Rev. Phytopathol., 3, 85
74. Mes-Hartree M and Saddler JN (1983) Biotechnol. Lett., 5(8), 531
75. Martin SA and Akin DE (1988) Appl. Environ. Microbiol., 54, 3019
76. Dekker RFH (1988) Appl. Environ. Biotechnol., 29, 593
77. Pfeifer PA, Bonn G and Bobleter O (1984) Biotechnol. Lett., 6(8), 541
78. du Preez JC, Bosch M and Prior BA (1986) Appl. Microbiol. Biotechnol., 23, 228
79. Skoog K and Hahn-Hägerdahl B (1988) Enzyme Microb. Technol., 10, 66
80. Tolan JS and Finn RK (1987) Appl. Environ. Microbiol., 53(9), 2033
81. Tolan JS and Finn RK (1987) Appl. Environ. Microbiol., 53(9), 2039
82. Chung IS and Lee YY (1986) Enzyme Microb. Technol., 8, 503
83. Delgenes JP, Moletta R and Navarro JM (1986) Biotechnol. Lett., 8(12), 897
84. du Preez JC, van Driessel B and Prior BA (1988) Biotechnol. Lett., 12, 901
85. Ligthelm ME, Prior BA and du Preez JC (1988) Biotechnol. Lett., 10(3), 207
86. Ligthelm ME, Prior BA and du Preez JC (1989) Biotechnol. Bioeng., 32, 839
87. Wayman M and Parekh S (1985) Biotechnol. Lett., 7(12), 909
88. Novak M, Strehaiano P, Moreno M and Goma G (1981) Biotechnol. Bioeng. 23, 201
89. Ando S, Arai I, Kiyoto K and Hanai S (1986) J. Ferment. Technol., 64(6), 567
90. Nishikawa NK, Sutcliffe R and Saddler JN (1988) Appl. Microb. Technol., 27, 549
91. Tran AV and Chambers RP (1985) Biotechnol. Lett., 7(11), 841
92. Dekker RFH (1986) Biotechnol. Bioeng., 28, 605

Application of Immobilized Growing Cells

Atuso Tanaka and Hiroki Nakajima
Laboratory of Industrial Biochemistry, Department of Industrial Chemistry,
Faculty of Engineering, Kyoto University, Yosida, Sakyo-ku, Kyoto 606, Japan

Immobilized living and growing cells are attracting worldwide attention because these biocatalysts have self-proliferating and self-regenerating properties of catalytic systems and are able to catalyze efficiently multifunctional and multistep reactions involving coenzyme regeneration. This article summarizes the application of microbial, plant, and mammalian cells, genetically improved or not, immobilized by different methods to the production of amino acids, organic acids, antibiotics, steroids, medicines, enzymes, bioactive peptides, etc., emphasizing the recent results. Effects of the gel properties on the efficient performance of bioprocesses are also discussed.

1 Introduction

Bioprocesses have received increasing worldwide attention with respect to low environmental pollution and economical utilization of natural resources and energy. They have also contributed to a wide variety of fields such as the production of useful substances and energy, artificial organs or slow-releasing drugs for medical purposes, degradation or removal of environmental contaminants, and analysis of various compounds with high specificity and high sensitivity. The

Advances in Biochemical Engineering/
Biotechnology, Vol. 42
Managing Editor: A. Fiechter
© Springer-Verlag Berlin Heidelberg 1990

construction of bioprocesses has been carried out with various kinds of biocatalysts, that is, single enzymes, multi-enzyme conjugating systems, cellular organelles, microbial cells, plant cells, and animal cells, immobilization of which has been demonstrated to possibly increase the efficiency of such bioprocesses. Studies on immobilized biocatalysts were initiated by immobilizing single enzymes for simple reactions such as hydrolysis and isomerization. Subsequently, multi-enzyme systems, isolated cellular organelles, and treated microbial cells have been used as biocatalysts for more complicated or conjugated reactions. Moreover, studies have been developed towards utilization of living or growing microbial cells and of multicellular organisms (cultured cells of higher plants and animals) as well as genetically improved microbial cells.

Among these immobilized biocatalysts, particularly the immobilized growing cells have recently attracted a great deal attention. The reasons for this are as follows:

(1) Growing or living cells possess numerous multi-enzyme systems catalyzing the biosynthesis of various useful substances. Growing cells are, therefore, able to synthesize metabolites with complicated structures through organized complex reactions, which can not be catalyzed by the combination of isolated enzymes.

(2) Immobilized growing cells are able to proliferate in or on support materials (carriers) and to regenerate catalytic systems in the cells by the supply of nutrients from culture media — self-proliferation and self-regeneration of catalytic systems. Consequently, catalytic activities of immobilized growing cells are expected to be maintained during long repeated or continuous of the cells.

(3) The density of immobilized growing cells in or on the carriers is locally higher than that of freely suspended counterparts after cell growth. This high cell density seems to be advantageous for reactions.

(4) Immobilized growing cells are easily separated from the reaction systems. Additionally, repeated or continuous use of immobilized growing cells is possible in various types of reactors.

(5) When cells are cultivated under poorly growing conditions to produce some secondary metabolites, wash-out of the cells from reactors can be avoided even at a high dilution rate.

On the other hand, immobilized growing cells have also some defects or problems as biocatalysts, and ways to overcome these disadvantages have been actively studied.

(1) Immobilized growing cells demand various nutrients and energy for proliferation or for preservation of viability. The energy efficiency to a target reaction is, therefore, apt to become low. In order to increase the energy efficiency, cell growth is occasionally restricted by limitation of nutrients while preserving a catalytic activity.

(2) Products are liable to be contaminated by cells leaking from the carriers. This problem will be solved by selection of a suitable carrier with a high cell-holding capacity and by limitation of excess cell proliferation at the carrier surface.

(3) By-products are ready to be synthesized because various metabolic systems in addition to the desired reactions are contained in living cells. Selection of a

potent-producing strain or genetic improvement of a cell line is essential to defeat this problem.

In spite of these drawbacks, studies on immobilized growing cells have been extensively carried out because of their advantages as biocatalysts, as mentioned above [1]. Industrial application of immobilized growing cells has, however, only reached the stage of pilot plant, development of the processes being expected in various fields.

2 Preparation of Immobilized Growing Cells

Numerous immobilization methods have been demonstrated hitherto, and the selection of a suitable method and a carrier for objective cells and reaction is of considerable importance in order to reach high performance by the immobilized growing cells. In general, it is difficult to predict an optimal carrier exactly, although the relationship between the efficiency of the reaction and the physico-chemical properties of each carrier is becoming increasingly clear. At present, the most suitable method and carrier should be, therefore, selected on a trial and error basis.

Immobilization methods can be classified into three categories, that is, carrier binding, cross-linking, and entrapping. The last method has been mainly used for the immobilization of living cells. Entrapment of living cells is achieved by using polymer gels, microcapsules, liposomes, hollow fibers, and ultrafiltration membranes. In particular, entrapment in polymer matrices (lattice type) seems to be suitable for the preparation of immobilized growing cells. For this purpose, various types of natural and synthetic polymers have been applied hitherto [2].

Entrapment of living cells with natural polymers, such as agar, agarose, alginate, ×-carrageenan, etc. is principally carried out by ionotropic or thermal gelation (Table 1). The formation of ionic networks is performed by dripping a cell/polymer

Table 1. Typical materials used for the entrapment of living cells

Material	Principle of gelation
Polysaccharides	
Alginate	Ionotropic gelation by multivalent cations
×-Carrageenan	Thermal gelation and ionotropic stabilization
Agar	Thermal gelation
Agarose	Thermal gelation
Synthetic chemicals	
Acrylamide	Polymerization with cross-linking reagent by addition of initiator and catalyst
Polyacrylamide-hydrazide	Polymerization with cross-linking reagent
Urethane resin prepolymers	Polycondensation in the presence of water
Photo-crosslinkable resin prepolymers	Radical-mediated polymerization by irradiation with near ultra-violet light
Photosensitive resin prepolymers	Cross-linking through photo-dimerization by irradiation with visible light

suspension into a solution containing a multivalent cation, and thermal gelation is achieved by cooling of a heated cell/polymer suspension. Since only nontoxic reagents are used under mild conditions in these procedures, damage to living cells during the process of immobilization can be minimized. Moreover, natural polysaccharides sometimes elicit synthesis of specific secondary metabolites in plant cells. These methods have been, therefore, preferentially chosen for the immobilization of living cells. However, multivalent cations or preheating of a cell/polymer suspension sometimes have an unfavorable effect on cell viability and productivity of secondary metabolites. Furthermore, some disadvantages such as mechanical strength and longevity of the gel structure have prevented natural polymers from being applied in industrial procedures before adequate reinforcement of the gel structure. On the contrary, the application of synthetic polymers to immobilization of living cells attracts great interest in transforming a wide variety of substances differing in nature, since synthetic polymers of adequate properties can be easily and artifically provided by designing their physicochemical structures depending on the reactions requested. Additionally, mechanical strength and longevity of the gels formed from synthetic polymers are generally superior to those from natural polymers.

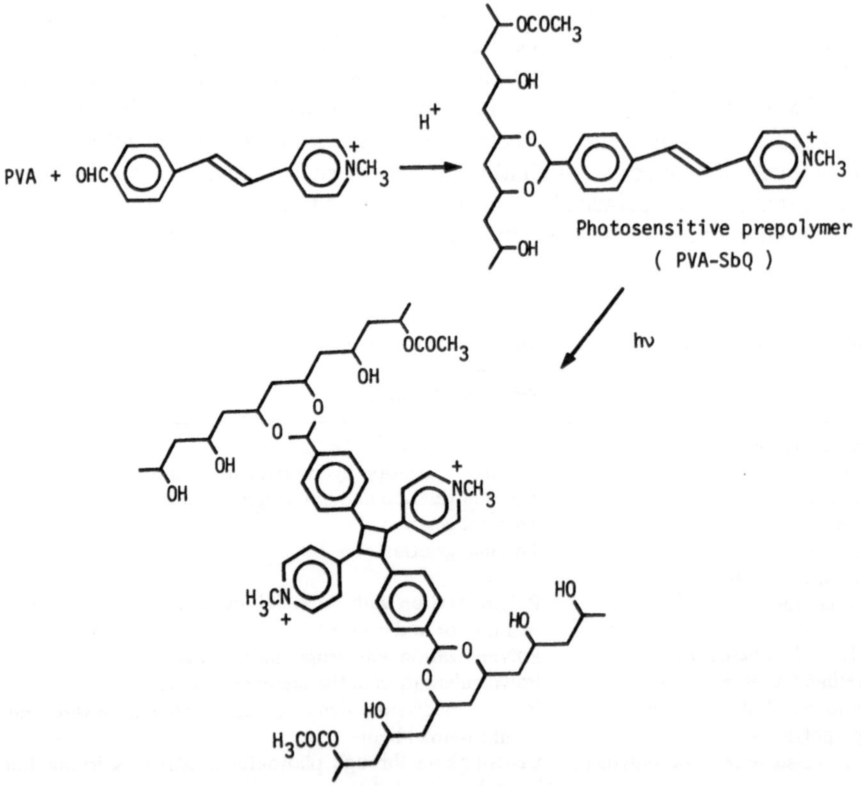

Fig. 1. Structure and photo-dimerization of photosensitive resin prepolymer (PVA-SbQ)

Acrylamide gels are often used for the entrapment of living cells as a representative synthetic polymer [3]. However, this procedure sometimes damages cell viability because of the toxicity of chemicals used for the gel preparation. To eliminate unfavorable influences caused by the acrylamide monomer, prepolymerized acrylamide derivatives (polyacrylamide-hydrazide) were devised as the gel material for the entrapment of living cells [4].

Simple and convenient methods have been developed for the entrapment of biocatalysts by using photo-crosslinkable resin prepolymers [5] and urethane prepolymers [6], and these prepolymers have been applied to the immobilization of various biocatalysts [7, 8]. The size of the network of the gel matrices, which affects diffusion of substrate and product inside the gel, mechanical strength of the gel formed, growth ability of cells inside gel matrices, or cell-holding capacity of the gel, can be regulated by the chain lengths of the prepolymers and content of the reactive functional groups. Furthermore, a balance between hydrophilicity and hydrophobicity can be optionally controlled by selecting a prepolymer or by mixing prepolymers of different properties at a proper ratio. The hydrophilicity-hydrophobicity balance of the gel often gives critical effects on the reaction rate by affecting the partition of reactants to the gel. Ionic characters of the gels, which are also an important factor to carry out enzymatic reactions efficiently, are able to be introduced to the prepolymers by using ionic oligomers as the main chain.

Photosensitive resin prepolymers (Fig. 1), which have been developed by Ichimura [9], are derivatives of poly(vinyl alcohol) introduced by styrylpyrydinium groups as photosensitive functional sites and are polymerized by photo-dimerization with irradation of visible or ultra-violet light. Hydrophilicity of the prepolymers can be controlled by changing the saponification degree of poly(vinyl alcohol). In the meantime, Kumakura et al. [10] reported the entrapment of living cells with 2-hydroxyethyl methacrylate, 2-hydroxyethyl acrylate, and others under γ-ray irradiation at low temperatures such as −78 °C.

Adsorption of cells on suitable carriers is also a potent alternative immobilization method of industrial importance.

3 Application of Immobilized Growing Cells

3.1 Immobilized Growing Microbial Cells

Microbial cells are thought to be excellent biocatalysts because of a high growth rate, insensitivity against shear force, and a relatively low cost for cultivation. Moreover, microbial cells can synthesize and secrete various secondary metabolites as well as primary metabolites. Immobilized growing microbial cells have been, therefore, utilized for production and biotransformation purposes to yield various useful substances, such as amino acids, organic acids, antibiotics, steroids, enzymes, peptides, etc. Processes using immobilized growing microbial cells seem to be more promising than traditional microbial processes with free cells since the immobilization enables the microbial cells to be used repeatedly or continuously.

3.1.1 Production of Amino Acids

Amino acids are widely used for medical purposes and as additives of foods, feeds, and cosmetics. L-Isomers of amino acids are mainly applied for these purposes, although D-isomers are useful for the synthesis of antibiotics. Biosynthesis of L-amino acids by microbial cells and optical resolution of chemically synthesized amino acids by microbial enzymes have been extensively investigated. Several processes have been successfully applied on an industrial scale, in which immobilized treated microbial cells are employed to catalyze single enzymatic reactions. In the meantime, several studies have been reported concerning the production of amino acids by immobilized growing microbial cells as summarized in Table 2.

Growing cells of *Serratia marcescens* immobilized with x-carrageenan were used for the continuous production of L-isoleucine in a multi-fluidized-bed reactor [13]. The immobilized cells exhibited a stable isoleucine-producing activity when the pH of the medium was controlled automatically at 7.5. In a two-bed reactor system, the L-isoleucine concentration of the effluent reached 4.5 mg ml^{-1} after the incubation for 10 h, and a steady production state was maintained for more than 30 days.

Continuous production of L-arginine was also carried out by growing *S. marcescens* cells immobilized with x-carrageenan [15]. Gel-entrapped *S. marcescens* cells were in an oxygen-limited state due to the diffusion barrier for oxygen transport constituted by the gel layer, so that the productivity of the amino acid was rather low. In order to increase the productivity, oxygen-enriched gas instead of air was supplied into the medium. Consequently, more than 10 mg ml^{-1} of L-arginine was continuously produced for at least 12 days under constant control at pH 6.5. L-Serine was also produced continuously by calcium alginate-entrapped L-leucine-auxotrophic cells of *Corynebacterium glycinophilum* at a concentration of 5–6 mg ml^{-1} [16].

Immobilized growing cells of *Corynebacterium glutamicum* prepared with acrylamide were employed for the production of L-glutamic acid in successive batches [11]. The immobilized growing cells produced more glutamic acid than the free cells under the same conditions. Living cells of *Brevibacterium flavum*

Table 2. Production of amino acids by immobilized growing microbial cells

Amino acid	Microorganism	Gel material	Ref.
L-Glutamic acid	*Corynebacterium glutamicum*	Polyacrylamide	[11]
	Brevibacterium flavum	Collagen	[12]
L-Isoleucine	*Serratia marcescens*	x-Carrageenan	[13]
L-Alanine	*Corynebacterium dismutans*	Polyacrylamide, x-carrageenan, DEAE-cellulose	[14]
L-Arginine	*Serratia marcescens*	x-Carrageenan	[15]
L-Serine	*Corynebacterium glycinophilum*	Calcium alginate	[16]

immobilized in collagen were used for the continuous production of L-glutamic acid in a recycle reactor system [12].

3.1.2 Production of Organic Acids

Organic acids are important microbial products used as food additives and medicines. Industrial processes for the production of organic acids have been carried out by using immobilized treated microbial cells as functional catalysts similarly to those used for the production of amino acids. Many studies on the production of organic acids by immobilized growing microbial cells also have been performed.

One group of organic acids including citric acid, itaconic acid, and lactic acid can be synthesized *de novo* from carbohydrates by immobilized growing bacterial or fungal cells (Table 3).

Production of citric acid from sucrose has been examined by immobilized growing cells of *Aspergillus niger*. Vaija et al. [17] employed an air-lift completely stirred reactor with calcium alginate-entrapped cells for the continuous production of citric acid, the maximum rate of which was 70 mg g^{-1} h^{-1} with an efficiency of 40%. Horitsu et al. [19] carried out continuous production of citric acid with polyacrylamide-entrapped growing cells in a two stage bioreactor. In this examination, the maximum production rate reached 96.6 mg h^{-1} per 80-g gels after 8 days, and the half-life period was found to be approximate 96 days. A productivity of 0.135 g l^{-1} h^{-1} was attained with the fungal cells immobilized by adsorption on polyurethane foam [20].

Itaconic acid, which is used to manufacture polyester resin and plastics, can be synthesized from glucose by immobilized growing cells of *Aspergillus terreus*. The growing cells immobilized in polyacrylamide were able to produce continuously itaconic acid with a maximum rate of 60 mg h^{-1} per 40-g gels in a column

Table 3. Production of organic acids by immobilized growing microbial cells

Organic acid	Microorganism	Gel material	Ref.
Citric acid	*Aspergillus niger*	Calcium alginate	[17]
		ϰ-Carrageenan	[18]
		Polyacrylamide	[19]
		Polyurethane foam	[20]
Itaconic acid	*Aspergillus terreus*	Polyacrylamide	[21]
		Calcium alginate	[22]
		Polyurethane foam	[23]
Lactic acid	*Lactobacillus delbrueckii*	Calcium alginate	[24]
		Hollow fiber	[25]
	Lactobacillus helveticus	Calcium alginate	[26]
	Lactobacillus vaccinostercus	Calcium alginate	[27]
	Rhizopus oryzae	Calcium alginate	[28]
Gluconic acid	*Aspergillus niger*	Glass carrier	[29]
2-Ketogluconic acid	*Serratia marcescens*	Collagen	[30]

Table 4. Production of organic acids via bioconversion by immobilized growing microbial cells

Organic acid	Substrate	Microorganism	Gel material	Ref.
Acetic acid	Ethanol	*Acetobacter* sp.	Hydrous titanium(IV) oxide	[31]
			Polypropylene fiber	[32]
		Acetobacter aceti	Ceramic monolith	[33, 34, 35]
			Ceramics	[36]
			х-Carrageenan	[37, 38]
		Acetobacter rancens	Hollow fiber	[39]
α-Keto-γ-methiol-butyric acid	L-Methionine	*Trigonopsis variabilis*	Calcium alginate	[40]
α-Keto-γ-methiol-butyric acid	D-Methionine	*Providencia* sp.	Calcium alginate	[41]
α-Keto-isocapric acid	L-Leucine	*Providencia* sp. + *Chlorella vulgaris*	Agarose	[42]
α,ω-Alkanedioic acid	n-Alkanediol	*Candida tropicalis*	х-Carrageenan	[43]

reactor [21]. The fungal cells adsorbed on polyurethane foam accumulated 13.3 g l^{-1} of the acid [23].

Production of lactic acid by growing cells of *Lactobacillus delbrueckii* entrapped in calcium alginate [24] and in hollow fibers [25] has also been reported.

Other types of organic acids have been synthesized by oxidation of amino acids or alcohols by immobilized growing cells (Table 4). α-Keto acids, which are useful for the treatment of chronic uremia, can be synthesized from the corresponding amino acids by microbial cells having amino acid oxidase activities.

Trigonopsis variabilis cells having a high level of D-amino acid oxidase were entrapped in calcium alginate containing manganese oxide and employed for the production of α-keto-γ-methiolbutyrate from the racemic mixture of methionine with simultaneous optical resolution. Co-immobilized manganese oxide efficiently degraded toxic hydrogen peroxide formed as one of the products and simultaneously generated oxygen necessary for the enzymatic reaction [40]. In the case of calcium alginate-entrapped *Providencia* sp. cells containing L-amino acid oxidase, co-immobilized activated charcoal also played effective roles in degrading hydrogen peroxide and supplying oxygen [41]. Furthermore, agarose-entrapped *Providencia* sp. cells showed an increased production of α-ketoisocapric acid by co-immobilization with *Chlorella vulgaris* cells under illumination of red light, so that photosynthesis by algal cells improved oxygen supply to the immobilized microbial cells [42].

Acetic acid is produced from ethanol by acetic acid bacteria at low pH associated with cell growth. Several groups have investigated the production of acetic acid by growing cells of *Acetobacter* sp. immobilized with hydrous titanium [31], ceramic monolith [33, 34, 35], hollow fibers [39], cotton-like fibers [32], and х-carrageenan [37, 38]. In the case of growing *Acetobacter aceti* cells immobilized with х-carrageenan, a steady production was achieved at the rate of about

$4 \text{ mg ml}^{-1} \text{ h}^{-1}$ for 120 days with aeration of pure oxygen gas, the supply of which was the limiting factor for increasing the cell number and the production activity.

Formation of dicarboxylic acid from n-alkanediol by *Candida tropicalis* cells entrapped with ×-carrageenan was also reported [43].

3.1.3 Continuous Production of Antibiotics

Production of antibiotics, which is one of the most important subjects in the field of biochemical engineering, has been carried out through microbial processes, enzymatic reactions, chemical synthesis, or combinations of these methods. Microbial processes mainly have been performed with batch-wise systems hitherto, because antibiotics are synthesized *de novo* after exponential growth of microbial cells, that is, antibiotics are non-growth-associated secondary metabolites, and the producing activities of microorganisms are often unstable. It is, therefore, quite difficult to produce antibiotics continuously during the prolonged cultivation of microbial cells. Furthermore, mycelium-forming microbial cells, such as actinomycetes or fungi, are employed for the production of antibiotics, so that it is not easy to construct continuous production systems. These problems are, however, overcome to some extent by using immobilized microbial cells. Oxygen supply is one of the limiting factors for the cultivation of immobilized cells. Co-

Table 5. Production of antibiotics by immobilized growing microbial cells

Antibiotic	Microorganism	Gel material	Ref.
Penicillin G	*Penicillium chrysogenum*	Polyacrylamide	[44, 45*]
C^{14}-Penicillin G	*Penicillium chrysogenum*	Calcium alginate	[46]
Cephalosporins	*Streptomyces clavuligerus*	Polyacrylamide-hydrazide	[47]
β-Lactam antibiotics	*Cephalosporium acremonium*	Calcium alginate	[48]
	Cephalosporium acremonium + *Chlorella pyrenoidosa*	Calcium alginate	[49]
Bacitracin	*Bacillus* sp.	Polyacrylamide	[50*]
Nikkomycin	*Streptomyces tendae*	Calcium alginate	[51*]
Tylosin	*Streptomyces* sp.	Calcium alginate	[51*]
Cyclosporin A	*Tolypocladium inflatum*	×-Carrageenan	[52]
Thienamycin	*Streptomyces cattleya*	Celite	[53]
Patulin	*Penicillium urticae*	×-Carrageenan	[54, 55*, 56*]
Oxytetracycline	*Streptomyces rimosus*	Polyurethane	[57*]
Daunorubicin	*Streptomyces peucetius*	Calcium alginate, photosensitive resin	[58]
Actinomycin D	*Streptomyces parvullus*	Calcium alginate	[59]

* Continuous production

immobilization of an oxygen-generating alga with an oxygen-consuming organism may be effective [49] as well as introduction of oxygen-rich air and improvement of reactors.

Continuous production of antibiotics seems to be feasible with immobilized growing cells due to maintenance of a high cell density in the reactor and relatively easy control of reactor operation, as in the case of the production of bacitracin [50], patulin [54, 55, 56], thienamycin [53], nikkomycin [51], tylosin [51], and penicillin G [45] as summarized in Table 5. Several investigations on the batch-wise production of antibiotics as a pre-stage of continuous operation have also been carried out.

The authors have demonstrated the advantageous use of immobilized growing cells for the continuous production of oxytetracycline by *Streptomyces rimosus* [57]. Immobilized growing cells of *S. rimosus* prepared with various gel-forming materials synthesized *de novo* oxytetracycline after exponential growth phase. Among these materials, a hydrophilic urethane prepolymer was found to be superior to the others with respect to the oxytetracycline productivity of the entrapped cells, cell-holding capacity, and mechanical strength of the gel. Generally, a great share of nutrients are consumed for the cell proliferation when abundant nutrients are available through primary metabolism. Production of non-growth-associated secondary metabolites, such as antibiotics, has been, therefore, carried out in media containing limited nutrients. Use of a glucose-free medium for the cultivation of the entrapped cells increased the production rate of oxytetracycline and minimized cell leakage from the gels. Such limitation of nutrients was also reported to be effective for the production of bacitracin [50] and patulin [56], where 0.5 % peptone solution and a nitrogen source-limited medium, respectively, were used for the continuous cultivation. Immobilization was found to be effective to avoid the wash-out of the poorly growing cells from the reactor.

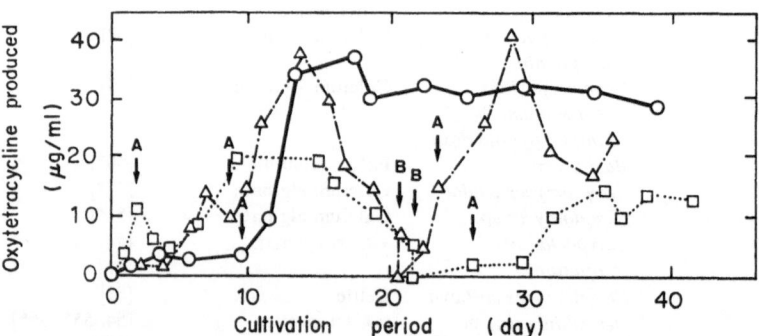

Fig. 2. Continuous production of oxytetracycline in different types of air bubbled column reactors with polyurethane-entrapped growing *Streptomyces rimosus* cells.
(○), Draft tube reactor (working volume 190 ml; 5.6 g gel);
(△), fixed type reactor (working volume 180 ml; 8.0 g gel);
(□), moving type reactor (working volume 150 ml; 10.0 g gel).
A, start of continuous cultivation; **B**, wash with saline for 2 days

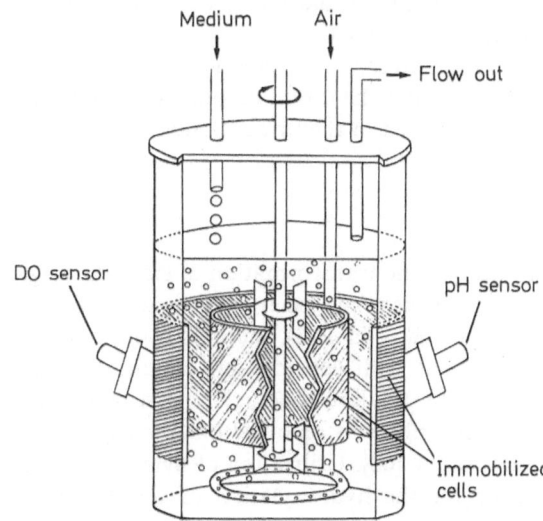

Medium   Air

→ Flow out

DO sensor

pH sensor

Immobilized
cells

Fig. 3. Agitated bubble column reactor used for continuous production of oxytetracycline by polyurethane-entrapped growing *Streptomyces rimosus* cells. Working volume: 2 l

When the entrapped cells lost the activity, recovery of the oxytetracycline productivity was observed by treatment of the cell-entrapping gels with saline or 70% ethanol to remove inactivated cells from the gel surface. A similar effect was shown in the production of bacitracin [50]. Continuous oxytetracycline production using polyurethane-entrapped growing cells was successfully achieved in an air-bubbled reactor for at least 35 days with the intermittent reactivation of the entrapped cells. Moreover, continuous production of oxytetracycline by the polyurethane gel-entrapped growing cells was compared in various operation types of reactors, and the oxytetracycline productivity could be stabilized for more than 50 days at a high level (30 μg ml^{-1}) in a draft tube-type reactor (Fig. 2), which is a type of bubble-column reactor (180 ml) containing a cell-entrapping gel film. Based on these results, increased production of oxytetracycline was successfully achieved in an agitated bubble-column reactor (Fig. 3) scaled up 10 times (Ogaki et al., unpublished data).

Streptomyces peucetius cells immobilized with calcium alginate and a photosensitive resin produced daunorubicin (daunomycin), an antibiotic which is clinically useful as a powerful inducer of remissions in acute leukemias, in modified media which had been optimized to prevent its rapid degradation. These entrapped cells were able to be used for the repeated daunorubicin production at least five times for 45 days [58].

Recently, drug-delivery systems have been developed by using microcapsule- or liposome-entrapped medicines. The authors have attempted to develop a slow drug-releasing system with immobilized living microbial cells implanted into an animal body, the immobilized cells being able to produce and release a drug continuously into the animal's body under the symbiosis with surrounding animal tissues. As a model, polyurethane-entrapped *S. rimosus* cells were implanted in rabbits and mice for oxytetracycline release from the entrapped cells to the animal's bodies. The polyurethane gel showed good compatibility with the surrounding tissues, and a sufficient but not over proliferation of the entrapped cells was observed in the gel matrices. A compound which had an identical retention time as oxytetracycline on HPLC was successfully secreted into the rabbit blood, and an antibiotic activity similar to oxytetracycline was detected in the urine (Sonomoto et al., unpublished data).

3.1.4 Transformation of Steroids

Various microbial cells are able to catalyze the transformation of steroids. Regio- and stereo-specific hydroxylation of steroids has been investigated by using immobilized growing or living microbial cells (Table 6). Steroids hydroxylated at a desired position are useful raw materials with considerable commercial value for the production of pharmaceutical steroid hormones. The first example of steroid conversion by immobilized living cells was reported on Δ^1-dehydrogenation of hydrocortisone to yield prednisolone [68]. Utilization of living or growing cells is supposed to be advantageous for the hydroxylation of steroids, which involves quite complex reactions including activation of molecular oxygen and continuous supply of reducing power.

Hydroxylation of steroids is often catalyzed by fungi, the fully developed mycelia of which generally possess a high activity. On the contrary, mycelia homogenized to prepare entrapped cells almost completely lose their hydroxylation activity. A method, in which fungal spores are entrapped homogeneously and permitted to germinate and to develop mycelia inside the gel matrices by cultivation in a nutrient medium, has been devised in order to prepare gel-entrapped mycelia having a high hydroxylation activity [60]. The gel-entrapped spores of *Curvularia lunata* have been shown to grow into mycelia in gels prepared from several synthetic prepolymers and natural polysaccharides, the immobilized living mycelia so prepared having a high steroid 11 β-hydroxylation activity [61]. Cortexolone (Reichstein's Compound S) was hydroxylated at the 11 β-position

Table 6. Transformation of steroids by immobilized living microbial cells

Substrate	Reaction	Microorganism	Gel material	Ref.
Cortexolone	11β-Hydroxylation	*Curvularia lunata*	Calcium alginate	[60]
			Photo-crosslinked resin	[61, 62]
Progesterone	11α-Hydroxylation	*Rhizopus nigricans*	Agar	[63]
		Rhizopus stolonifer	Photo-crosslinked resin	[64]
4-Androstene-3,17-dione	9α-Hydroxylation	*Corynebacterium* sp.	Photo-crosslinked resin	[65]
Dehydroepi-androsterone	16α-Hydroxylation	*Streptomyces roseochromogenes*	Photo-crosslinked resin	[66]
Esterone	16α-Hydroxylation	*Sepedonium ampullosporum*	Photo-crosslinked resin	[67]
Hydrocortisone	Δ^1-Dehydrogenation	*Arthrobacter simplex*	Calcium alginate	[68, 69, 70]

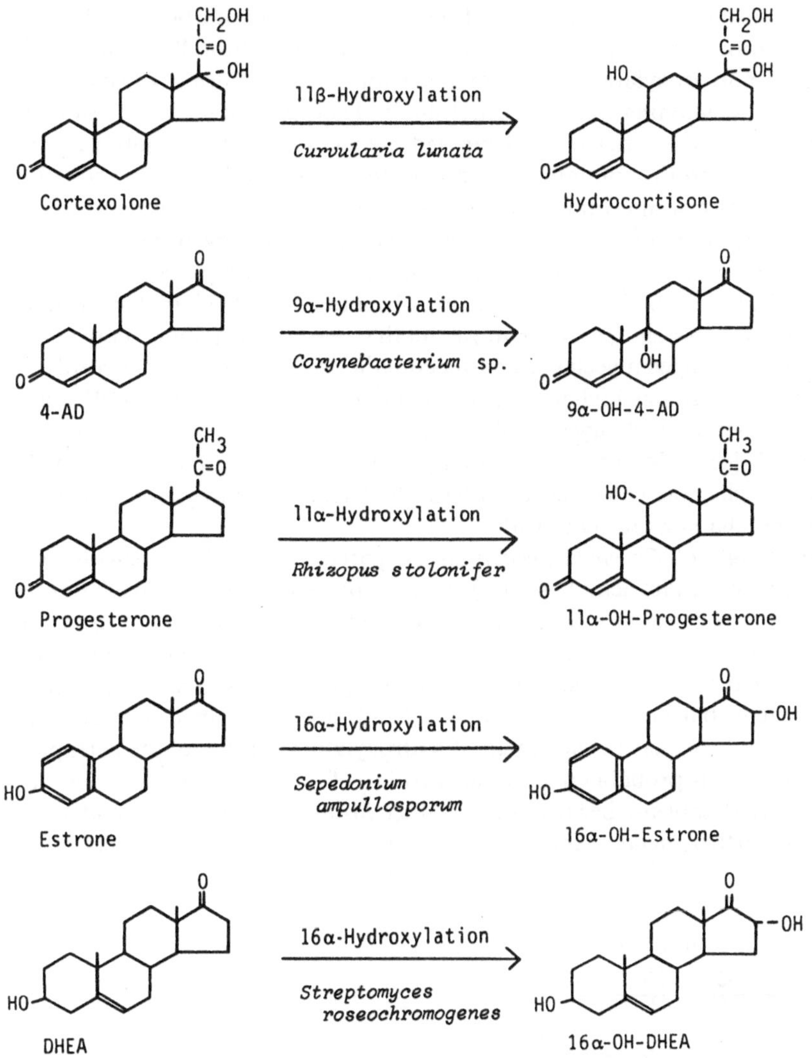

Fig. 4. Steroid hydroxylation catalyzed by immobilized microbial cells. 4-AD, 4-androstene-3,17-dione; DHEA, dehydroepiandrosterone

to form hydrocortisone by photo-crosslinked resin-entrapped mycelia of *C. lunata* (Fig. 4), which showed the highest hydroxylation activity among various entrapped cells tested. The chain length of the photo-crosslinkable resin prepolymers, that is, the network size of the gels formed had a significant effect on the development of the mycelia. Growth inside the gels prepared from the prepolymers having a shorter chain length was not sufficient, but growth inside the gels prepared from the longer chain prepolymers was good or moderate. Leakage of the mycelia from gels prepared from the long-chain prepolymers was, however, observed when the mycelium-entrapped gels were washed with buffer solutions. Mycelia entrapped with a prepolymer having a chain length of 40 nm were able to be used for hydroxylation of cortexolone for 50 times during 100 days in a buffer solution containing 2.5 % dimethyl sulfoxide or methanol with intermittent reactivation, when the activity was decreased, by incubating the immobilized mycelia in a nutrient broth containing cortexolone [62]. The possibility of reactivation is one of a number of important advantages of immobilized living cells.

11 α-Hydroxylation of progesterone was also examined by using entrapped living *Rhizopus stolonifer* mycelia [64]. Of various gel-forming materials tested, a mixture of photo-crosslinkable resin prepolymers (those having chain lengths of 20 and 40 nm) was chosen as the most suitable one. The maximum conversion obtained with entrapped mycelia in a buffer containing 2.5 % methanol as cosolvent was 73 %. This value was higher than that obtained with free mycelia (19 %) and that with agar-entrapped mycelia of *Rhizopus nigricans* (50 %) as reported by Maddox et al. [63]. Moreover, the stability of the hydroxylation activity in the entrapped mycelia was enhanced by intermittent reactivation by incubating the entrapped mycelia in a nutrient broth.

Entrapped cells of *Corynebacterium* sp. prepared with a photo-crosslinkable resin prepolymer (chain length 40 nm) were employed for 9α-hydroxylation of 4-androstene-3,17-dione in a buffer solution containing 15 % dimethyl sulfoxide, which was effective for the solubilization of the product and also enhanced the conversion yield [65]. The hydrophilicity of the photo-crosslinked resins in an aqueous system, but not in an organic cosolvent system, affected markedly the yield of the product, that is, the apparent hydroxylation activity of the entrapped cells (Fig. 5). This is probably due to the partition of the product between the gels and the external solvent, because the product inside the gels was found to be subjected to further metabolism by the cells. In the presence of nutrients in the

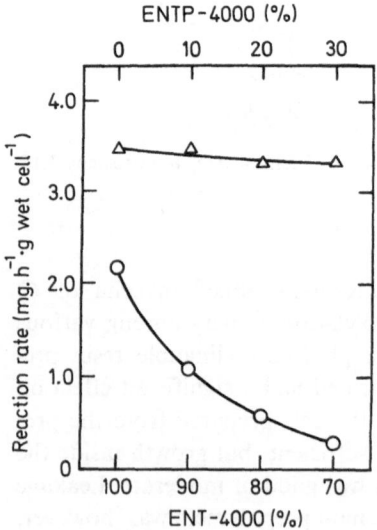

Fig. 5. Effect of gel hydrophobicity on apparent 9α-hydroxylation activity of *Corynebacterium* sp. cells entrapped with photo-crosslinkable resin prepolymers [65]. Cells were entrapped with various proportions of ENT-4000 (a hydrophilic prepolymer) and ENTP-4000 (a hydrophobic prepolymer).

($\bigcirc$), Reaction in 100 mM potassium phosphate buffer (pH 8.0);

($\triangle$), reaction in the buffer containing 15 % dimethyl sulfoxide reaction mixture, the entrapped growing cells maintainted fully the original activity for 5 days during 10 repeated batch reactions.

3.1.5 Production of Enzymes

Microbial cells are the best sources supplying large quantities of useful enzymes at a low price, and the production of extracellular enzymes, such as carbohydrate-hydrolyzing enzymes and proteolytic enzymes, has been mainly studied by using immobilized growing microbial cells (Table 7).

Amylases, which are among the most popular extracellular enzymes synthesized by *Bacillus* sp., are classified into several kinds according to the types of starch-hydrolysis, such as α-amylase and glucoamylase. Research on α-amylase production by polyacrylamide gel-entrapped *Bacillus* sp. cells was performed by Kokubu et al. [71]. This was the first example of enzyme production by immobilized growing microbial cells. The amount of α-amylase produced by immobilized whole cells of *Bacillus subtilis* was approximately three times higher than that produced by free cells under optimized conditions. The activity of the immobilized cells increased

Table 7. Production of enzymes by immobilized growing microbial cells

Enzyme	Microorganism	Gel material	Ref.
α-Amylase	*Bacillus subtilis*	Polyacrylamide	[71]
		×-Carrageenan	[72]
	Bacillus amyloliquefaciens	×-Carrageenan	[73]
		Ion exchange resin	[74]
	Bacillus subtilis + *Scenedesmus obliquus*	×-Carrageenan	[75]
Glucoamylase, α-amylase	*Aspergillus niger*	Calcium alginate	[76]
Cellulase	*Trichoderma reesei*	×-Carrageenan	[77]
		Nylon mesh	[78]
		Poly(2-hydroxyethyl methacrylate), poly(2-hydroxyethyl acrylate)	[10, 79]
		Polymerized 2,6-dimethylphenol (Sorfix)	[80]
	Talaromyces emersonii	Calcium alginate	[81]
Protease	*Streptomyces fradiae*	Polyacrylamide	[82]
	Serratia marcescens, *Myxococcus xanthus*	Calcium alginate	[83]
Acid protease	*Humicola lutea*	Polyacrylamide	[84]
Proteolytic enzyme	*Myxococcus xanthus*	×-Carrageenan	[85]
Leulysin	*Saccharomyces cerevisiae*	Photo-crosslinked resin	[86]
β-Lactamase	*Escherichia coli*	Hollow fiber	[87]
Lignin peroxidase	*Phanerochaete chrysosporium*	Nylon-web	[88]

gradually upon repeated use and reached a steady state after the seventh batch. On the other hand, free cells lost the activity during several successive batches. *Bacillus amyloliquefaciens* cells immobilized in x-carrageenan were employed for α-amylase production, the maximal rate of which was obtained under continuous cultivation at a dilution rate of 1.0 h^{-1} in an aerated vessel with a volume ratio of gel beads to medium of $1:2$ [73]. Similarly, α-amylase production was tested by using entrapped cells of *B. subtilis* [72]. Use of an airlift reactor with sparging air gave a 40–70% increase in α-amylase production over that of a round flask with bubbled air. *Aspergillus niger* immobilized in calcium alginate beads was also used for simultaneous production of glucoamylase and α-amylase by a repeated batch process [76].

Conversion of cellulosic biomass through enzymatic hydrolysis into mono- and oligosaccharides to be utilized by microbial cells has inherent economic problems, one of which is the cost of the preparation of cellulase. Several investigations have been attempted to reduce the cost of cellulase production by using immobilized growing cellulolytic cells of *Trichoderma reesei*. Continuous cultivation was carried out with *T. reesei* cells immobilized with x-carrageenan [77] and those held in a reactor by a nylon mesh [78]. *T. reesei* cells entrapped in synthetic polymers prepared by γ-ray irradiation at a low temperature were also employed for the production of cellulase in a batch culture [10, 79]. Moreover, the production of proteolytic enzymes was studied with *Streptomyces fradiae* cells [82] and *Myxococcus xanthus* cells [83, 85], which were immobilized in polyacrylamide and x-carrageenan, or calcium alginate, respectively.

During the study of the production of a yeast sex pheromone, the α-mating factor, *Saccharomyces cerevisiae* cells entrapped with a neutral hydrophilic photo-crosslinkable resin prepolymer were found to excrete specifically into a culture medium a novel type of peptidase, which cleaved the Leu-Lys bond of the α-mating factor (Fig. 6). This novel Leu-Lys-specific peptidase named "Leulysin" is one of the proteolytic enzymes having α-mating factor-degrading activity and was extracellularly produced at a high level only by cells entrapped at a high cell concentration [86]. Such specific peptidases will be of use for processing various physiologically active proteins and peptides, and also for study of protein structures.

Recently, increasing attention has been focused on the production of foreign proteins and enzymes by using plasmid-harboring strains of microbial cells, and several feasible studies on this subject have been carried out. Growing *Escherichia coli* C600 cells harboring a plasmid immobilized in hollow fiber membranes were reported to produce plasmid-mediated β-lactamase, the production rate of which

$$\text{Trp}^1\text{-His}^2\text{-Trp}^3\text{-Leu}^4\text{-Gln}^5\text{-Leu}^6\text{-Lys}^7\text{-Pro}^8\text{-Gly}^9\text{-Gln}^{10}\text{-Pro}^{11}\text{-Met}^{12}\text{-Tyr}^{13}$$

$\uparrow$

Fig. 6. Structure of α-mating factor. The arrow indicates the site of cleavage by a novel peptidase, Leulysin

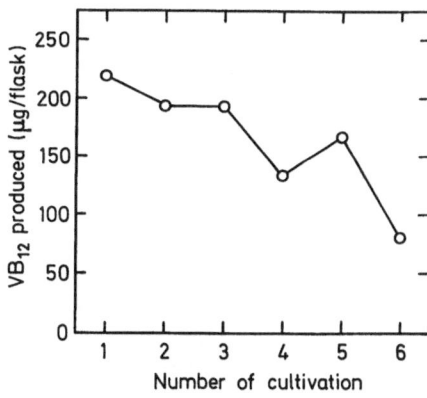

Fig. 7. Repeated use of immobilized cells of *Propionibacterium* sp. AKU 1251 for the production of vitamin B_{12} [89]
Cells were entrapped with a urethane prepolymer and each cultivation was carried out statically for 3 days in 50 ml of medium

remained at a relatively high and stable level for more than three weeks of continuous cultivation [87].

Besides the studies mentioned above, several examples on the production of enzymes by immobilized growing recombinant cells will be demonstrated later.

3.1.6 Production of Other Useful Substances

Immobilized growing microbial cells have been applied to the production of various organic compounds in addition to those described above.

Production of vitamin B_{12} was investigated by using immobilized *Propionibacterium* sp. cells. Whole cells were entrapped in various gel matrices such as polyurethane resins, photo-crosslinked resins, ϰ-carrageenan, agar, and calcium alginate. A hydrophilic urethane prepolymer was the most suitable material for vitamin B_{12} production. An increased productivity was obtained with the immobilized cells in the optimized medium supplemented with appropriate amounts of cobalt ions and 5,6-dimethylbenzimidazole as precursors. The immobilized cells retained the producing ability of vitamin B_{12} for several batches when 5 g of wet cells were entrapped with 1 g of the urethane prepolymer (Fig. 7), and around 1000 μg of vitamin B_{12} was excreted into the medium during 6 batches of cultivation (total medium volume, 300 ml) [89].

Production of the α-mating factor (a tridecapeptide, see Fig. 6) with immobilized growing *Saccharomyces cerevisiae* cells was studied in order to establish a secretion system for useful bioactive peptides using the α-mating factor secretion system [90]. The α-mating factor is a kind of microbial pheromone excreted into the culture medium by *S. cerevisiae* cells of the α-mating type. The α-mating factor thus excreted into the medium was reported to be degraded during cultivation by cleavage of its peptide bonds. When the yeast cells were entrapped in a mixture of neutral and anionic photo-crosslinkable resin prepolymers (Fig. 8), the activity of the α-mating factor in the medium was maintained for at least 5 days of cultivation after reaching the maximal level. On the other hand, secreted α-mating factor was degraded within a few days in the case of cells immobilized with other gel-forming materials tested and of the freely suspended cells. The reason for this

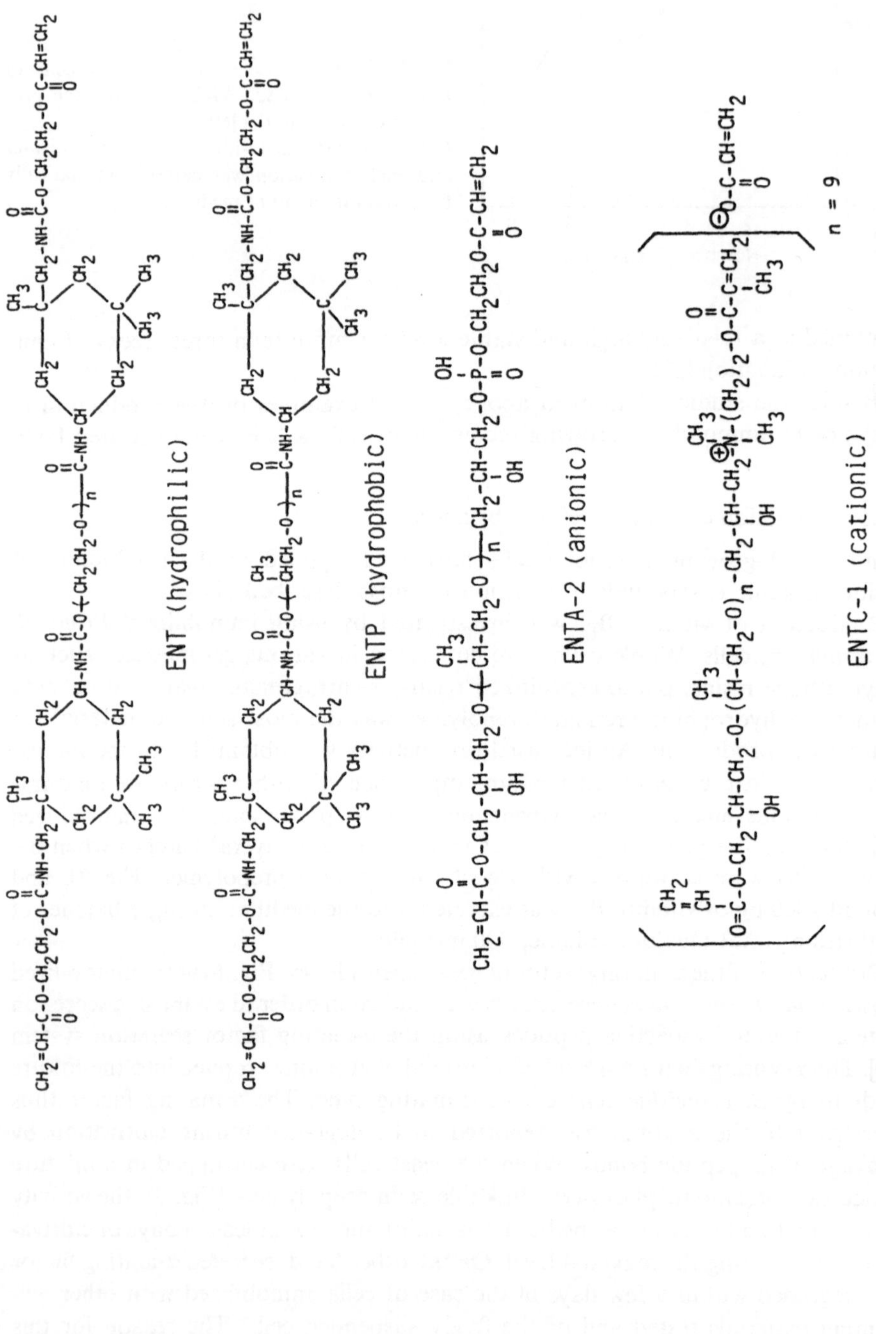

Fig. 8. Structures of typical photo-crosslinkable resin prepolymers. ENT, neutral and hydrophilic; ENTP, hydrophobic; ENTA-2, anionic; ENTC-1, cationic

phenomenon was that the anionic gel specifically entraps the enzymes showing the degradation activity toward α-mating factor. As a result, the growing yeast cells entrapped in the anionic gel could be used successfully for continuous production of α-mating factor in a packed column reactor for at least 30 days (Fig. 9).

In addition to the studies mentioned above, several examinations have been reported on the production of useful organic compounds such as bioactive peptides by immobilized, genetically improved microbial cells, which will be introduced in the section of recent topics below.

Microbial production of alcohols, such as ethanol, butanol, and isopropanol, has attracted a renewed interest in the field of the chemical synthetic industry due to a change in the economic feasibility as a consequence of high oil prices. A great number of papers have been published mainly on the production of ethanol, and large-scale production of ethanol by immobilized growing yeast cells has recently been attempted at the stage of pilot plant with yeast cells entrapped with photo-crosslinkable resin prepolymers or with a mixture of a prepolymer and alginate [91, 92, 93]. A suspended-bed reactor with a producing capacity of 5 kl day^{-1} of ethanol was constructed together with an apparatus to prepare a large volume of gel beads entrapping the cells.

Continuous isopropanol-butanol-ethanol production was studied by using immobilized *Clostridium beijerinckii* cells in a packed-bed reactor [94]. The productivity of the calcium alginate-entrapped cells in continuous use for 290 days was 3 to 4 times higher than that obtained in a batch process using free cells.

The production of hydrogen, which is attractive as a clean fuel resource, was attempted by *Clostridium butyricum* immobilized in polyacrylamide [95]. The immobilized cells continuously evolved hydrogen from glucose under aerobic conditions. Methane production from waste water by immobilized methanogenic bacteria was also reported [96].

Calcium alginate-entrapped cells of *Gibberella fujikuroi* were applied to the continuous production of gibberellic acid [97, 98].

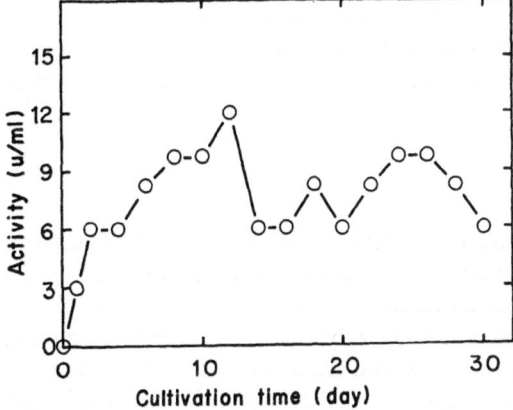

Fig. 9. Continuous production of α-mating factor by entrapped growing *Saccharomyces cerevisiae* cells [90]. Cells were entrapped with a mixture of ENT and ENTA-2 (8:2 by weight) and cultivated in a packed column reactor (working volume, 200 ml). Flow rate, 200 ml day^{-1}, aeration, 0.61 min^{-1}

3.2 Immobilized Plant Cells

Higher plants are important sources of numerous natural compounds such as pharmaceuticals, flavorings. coloring matters, which have complex structures and can not be synthesized by microbial cells. The production of these useful substances has been actively carried out by using cultured cells of higher plants, since original plants are quite disadvantageous as biocatalysts due to a low growth rate, unstable productivity, and changeable quality of products. Even cultured plant cells are, however, insufficient for the effective production of useful substances in respect of productivity, stability of production, and mechanical strength compared with microbial cells. Therefore, improvement of productivity by the techniques of cell engineering, such as selection of highly potent producing variants through cell-aggregate or protoplast cloning, optimization of culture conditions, protoplast fusion, and genetic engineering, has been attempted. In addition to these techniques, immobilization of cultured plant cells is suggested to be a simple and promising method from many investigations.

Immobilized cultured plant cells are supposed to have some merits concerning removal of disadvantages inherent to cultured plant cells as follows:

(1) Cultured plant cells are protected from external impact by entrapment with gel-forming materials.

(2) The problem of a low growth rate is relieved to some extent by immobilization of cultured plant cells at a high density and by repeated or continuous use of these immobilized cells.

(3) The productivity of secondary metabolites and product release from the cells sometimes increase by immobilization, since favorable stress is provided to cultured plant cells by immobilization. The materials for cell-entrapping can serve as elicitors in some cases for inducing the production of secondary metabolites.

(4) The products are easily separated and purified by preventing cell leakage from the gel under the limitation of cell proliferation.

(5) The forced aggregation of plant cells by immobilization will in some cases be effective for stabilizing cell viability and cell metabolism.

The application of immobilized plant cells is still a young field in biotechnology, and so the number of studies dealing with immobilized plant cells is much smaller than that concerning enzymes and microbial cells. Application of immobilized plant cells to the production of useful substances is grouped into three categories, that is, biotransformation, salvage synthesis, and *de novo* synthesis as described in the following sections.

3.2.1 Biotransformation

Immobilized plant cells have been utilized for several kinds of biotransformations which include stereo- and regio-specific hydroxylation, glycosylation, and reduction of organic compounds as summarized in Table 8.

Hydroxylation of cardiac glycosides and their aglycones was extensively studied by using immobilized cells of *Digitalis lanata* and *Daucus carota*. Brodelius et al. [99] reported that calcium alginate-entrapped *D. lanata* cells catalyzed 12β-

Table 8. Application of immobilized plant cells to the production of biochemicals

Plant cell	Support material	Substrate	Product	Ref.
Bioconversion				
Digitalis lanata	Calcium alginate	Digitoxin	Digoxin	[99]
		β-Methyldigitoxin	β-Methyldigoxin	[100]
		Digitoxin	Purpreaglycoside A	
Daucus carota	Calcium alginate	Digitoxigenin	5β-Hydroxydigitoxigenin (periplogenin)	[101]
Catharanthus roseus	Agarose	Gitoxigenin	5β-Hydroxygitoxigenin	[102]
	Polyacrylamide-hydrazide	Cathenamine	Ajmalicine isomers	[103]
Mentha sp.		(−)-Menthone	(+)-Neomenthol	[104]
Mucuna pruriens	Calcium alginate	(+)-Pulegone	(+)-Isomenthone	[105, 106]
		L-Tyrosine	L-Dihydroxyphenylalanine (L-DOPA)	
Papaver somniferum	Calcium alginate	Codeinone	Codeine	[107]
Salvage synthesis				
Catharanthus roseus	Calcium alginate, agarose, agar, x-carragenan	Tryptamine + Secologanin	Ajmalicine isomers	[99, 108, 109]
De novo synthesis				
Morinda citrifolia	Calcium alginate	Sucrose	Anthraquinones	[99]
Catharanthus roseus	Calcium alginate, agarose, agar, x-carragenan	Sucrose	Ajmalicine + serpentine	[110]
Glycine max	Hollow fiber	Sucrose	Phenolics	[111]
Daucus carota	Hollow fiber	Sucrose	Phenolics	[112]
Solanum aviculare	Polyphenyleneoxide	Sucrose	Steroid glycoalkaloids	[113]
Coffea arabica	Calcium alginate	Sucrose	Methylxanthine alkaloid	[114]
Capsicum frutescens	Polyurethane foam	Sucrose	Capsaicin	[115]
Solanum surattense	Calcium alginate	Sucrose	Solasodine	[116]
Lavandula vera	Calcium alginate, photosensitive resin	Sucrose, glucose	Blue pigments	[117, 118, 125]
Glycyrrhiza echinata	Calcium alginate	Sucrose	Echinatin	[119]
Salvia miltiorrhiza	Calcium alginate	Sucrose	Cryptotanshione + ferruginol	[120]
Thalictrum minus	Calcium alginate	Sucrose	Berberine	[121, 122]

hydroxylation of digitoxin to digoxin and that the catalytic activity of the immobilized cells was retained for more than 33 days. Alfermann et al. [100] demonstrated that immobilized cells of *D. lanata* prepared in calcium alginate gels preferentially converted digitoxin to purpureaglycoside A through glucosylation rather than 12β-hydroxylation. On the contrary, a 12β-hydroxylated product was obtained when β-methyldigitoxin was used as a substrate. The hydroxylation activity of calcium alginate-entrapped cells was one-and-a-half-fold higher than that of freely suspended cells under the same conditions, and the immobilized cells could be used for more than 60 days for the successive production of β-methyldigoxin, which has recently become clinically more useful than β-methyldigitoxin.

5β-Hydroxylation of digitoxigenin and gitoxigenin was studied by Jones and Veliky [101]. *D. carota* cells immobilized in calcium alginate produced periplogenin from digitoxigenin in five repeated batches with a designed buffer mixture. A rate greater than 60% of the original activity could be achieved before the immobilized cells were inactivated. When gitoxigenin was employed as a substrate, formation of 5β-hydroxygitoxigenin was observed in a column bioreactor and the conversion of 70–80% under upward aeration was maintained for 23 days with entrapped cells of *D. carota* [102].

Furuya et al. [107] investigated the biotransformation of an alkaloid codeinone to codeine by immobilized cells of *Papaver somniferum*. Cultured cells of *P. somniferum* were able to catalyze stereo- and regio-specific reduction of (−)-codeinone to (−)-codeine at a high yield, the reaction being one of the steps in *de novo* synthesis of morphine. When calcium alginate-entrapped cells were used in a shake flask, the conversion ratio for the administered codeinone was 70.4%. This value was higher than that obtained with the freely suspended cells (60.8%). Furthermore, 88% of the codeine converted by the immobilized cells was excreted into the medium. The conversion by the immobilized cells was achieved in a column bioreactor which was operated for 30 days. Under optimal conditions, the conversion ratio was 41.9%.

Mentha sp. cells immobilized in cross-linked polyacrylamide-hydrazide were used for the biotransformation of terpenoids by Galun et al. [104]. The conversion of (−)-menthone to (+)-neomenthol and of (+)-pulegone to (+)-isomenthone by the entrapped cells was as efficient as that obtained by the freely suspended cells. Less monoterpenes were found to be retained in the entrapped cells than in the case of the free counterparts.

In addition to the studies described above, Wichers et al. [105, 106] reported on the biotransformation of L-tyrosine to L-dihydroxyphenylalanine (L-DOPA) and *N*-formyl-tyrosine to *N*-formyl-DOPA by calcium alginate-entrapped cells of *Mucuna pruriens*. Agarose-entrapped cells of *Catharanthus roseus* were employed for the reduction of cathenamine to produce ajmalicine and its isomers [103].

3.2.2 Salvage Synthesis

Yields of secondary metabolites are supposed to be increased by feeding suitable precursors which are important intermediates on biosynthetic pathways of the target products to cultured plant cells. This production formula, called salvage synthesis, is considered to be an effective method for producing complex organic compounds. However, the application of salvage synthesis so far has been limited to the production of indole alkaloids (Table 8), since the information on the biosynthetic pathways of useful secondary metabolites is insufficient to apply this method, and proper precursors are not available in most cases.

· Tryptamine and secologanin are known to be the key intermediates in the biosynthesis of various indole alkaloids, and ajmalicine isomers could be produced from these precursors both by living cultured cells of *C. roseus* and by the cell-free extracts of the cultured cells. Brodelius et al. [99] investigated the production of ajmalicine isomers from tryptamine and secologanin by immobilized cells of *C. roseus*. The cells immobilized in calcium alginate were active in a solution containing the precursors in a recirculated packed bed reactor equipped with a vessel holding chloroform for the extraction of the lipophilic products. After incubation for 90 h, 30% of the tryptamine was converted to chloroform-extractable compounds, one third of which consisted of a mixture of ajmalicine and its isomers. A relatively high yield of the products was obtained by using immobilized cells. In freely suspended cells the products were not excreted into the medium but were stored within vacuoles. The observed excretion of the alkaloids formed is probably due to a change in the membrane permeability caused by traces of chloroform, which was used for the continuous extraction of the products. This product release caused by the permeabilization of membranes is considered to be of great importance because many secondary metabolites of plant cells are stored intracellularly.

Brodelius and Nilsson [108] further examined the salvage synthesis of ajmalicine isomers as a feasible study on the permeabilization of immobilized *C. roseus* cells to release intracellularly stored products while preserving the viability of the cells. When immobilized *C. roseus* cells were treated with medium containing 5% dimethyl sulfoxide, 85–90% of the intracellularly stored products, ajmalicine isomers, were released into the medium with retaining the cell viability. *C. roseus* cells entrapped in agarose and calcium alginate could be used for the repeated production of ajmalicine isomers in a cyclic process including intermittent permeabilization for release of the products, cell growth, and production.

3.2.3 *De novo* Synthesis

De novo synthesis of secondary metabolites specific to plant cells, such as alkaloids, essential oils, and pigments, has been actively studied. The contents of secondary metabolites in cultured plant cells are occasionally quite low since the *de novo* synthesis of secondary metabolites is closely associated with the functional differentation of plant cells. It is, therefore, difficult to obtain highly potent cell lines to produce target substances. Moreover, the production of secondary metabolites, which requires complex biosynthetic pathways, needs a relatively long duration of cultivation. As a result, studies on the *de novo* synthesis of useful substances by immobilized plant cells are not too numerous as shown in Table 8. However, the *de novo* synthesis of useful substances by immobilized plant cells is considered to become an economically efficient method depending on the productivity, since valuable products can be obtained from simple carbon sources.

Morinda citrifolia cells were entrapped in calcium alginate to investigate the production of anthraquinones [99]. The anthraquinone productivity of the immobilized cells increased during an extended cultivation in the absence of growth hormones, while that of the freely suspended cells decreased with time under the same conditions. After a cultivation for 22 days, the immobilized cells accumulated approximately ten times as much anthraquinones as the free cells.

De novo synthesis of ajmalicine was studied by using immobilized *C. roseus* cells prepared with various polysaccharide gel materials, such as agar, agarose, alginate, and ×-carrageenan [110]. All the preparations of the immobilized cells were able to synthesize *de novo* ajmalicine from sucrose at a rate of 60—40% of that observed with the freely suspended cells.

Lindsey et al. [115] reported the immobilization of plant cells on reticulated polyurethane foam and its application. Suspended cells of *Capsicum frutescens* and *Daucus carota* were found to be spontaneously retained on porous polyurethane foam particles and the viability of the immobilized cells was high. Polyurethane-adsorbed *C. frutescens* cells produced and released significantly more capsaicin into the medium than the freely suspended cells did. Addition of isocapric acid as a precursor enhanced five-fold the yield of capsaicin.

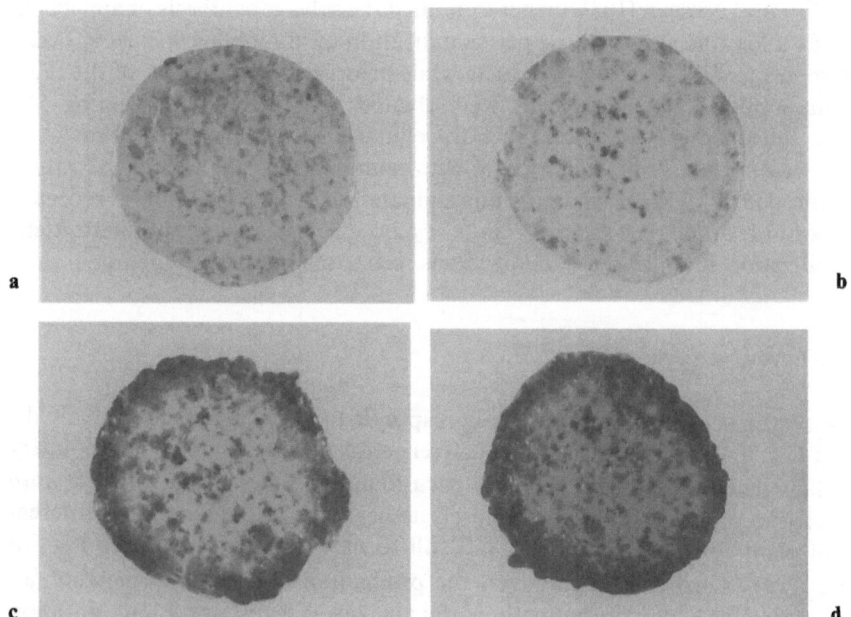

Fig. 10 a–d. Thin section of calcium alginate gel beads entrapping *Lavandula vera* cells [117]. Diameter of the beads is ca. 5 mm. **a** before cultivation; **b** cultivated for 8 days; **c** cultivated for 11 days; **d** cultivated for 15 days

Entrapped cells of *Glycine max* [111] and *Daucus carota* [112] in hollow fiber reactors were utilized for the synthesis of phenolics. A high level production (15–10 µg/ml) of phenolics was maintained for one month by the entrapped cells under the continuous operation of the reactors. Cultured cells of *Solanum aviculare* were immobilized by adsorption and covalent linkage on porous polyphenylene-oxide beads activated by glutaraldehyde and employed for the production of steroid glycoalkaloids in a recirculated packed column reactor. The immobilized cells could produce and release steroid glycoalkaloids during 11 days of incubation [113].

Production of diterpene cryptotanshione and ferrugiol was studied with immobilized cultured cells of *Salvia miltiorrhiza* [120]. The relative amounts of cryptotanshione and ferrugiol by the calcium alginate-entrapped cells were approximately 39% and 61%, respectively, of those produced by the freely suspended cells, but the immobilized cells were able to release higher amounts of cryptotanshione into the medium than did the freely suspended cells. Cultured cells of *Glycyrrhiza echinata* were reported to have the ability of rapid and transient accumulation of a retrochalcone echinatin in both the cells and the medium by transfer into a fresh medium or immobilization in calcium alginate, which induced higher and longer duration of echinatin production than did the other treatments. The gel-forming material, sodium alginate, was found to result in the induction of echinatin synthesis [119]. Berberine was produced extracellularly by calcium alginate gel-entrapped *Thalictrum minus* L. var. *hypoleucum* Miq. cells. Special reactors were devised for batch [121] and semicontinuous [122] production of berberine, because the supply of oxygen was found to be critical for the synthetic activity.

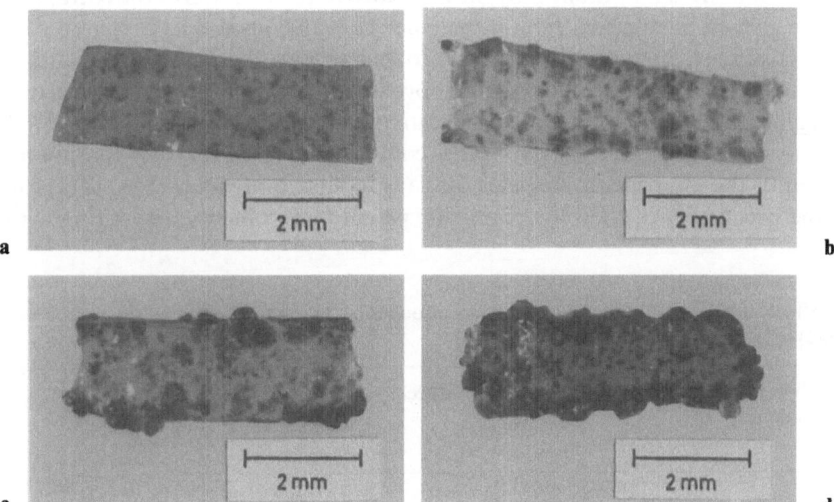

Fig. 11 a–d. Thin section of PVA-SbQ gel entrapping *Lavandula vera* cells [118]. The thickness of the gels is ca. 2 mm. **a** before cultivation; **b** cultivated for 8 days; **c** cultivated for 11 days; **d** cultivated for 15 days

Cultured cells of *Lavandula vera* strain L_{10} 4-2154 were found to synthesize *de novo* blue pigments whose formation was induced by the addition of L-cysteine to the medium [123]. Finely suspended cells of *L. vera*, which were obtained by subcultivation in a medium containing a relatively high concentration of calcium chloride, were immobilized with various kinds of gel-forming materials including agar, κ-carrageenan, alginate, urethane prepolymers, photo-crosslinkable resin prepolymers, and photosensitive resin prepolymers [117, 118]. Considering cell growth after immobilization, cell-holding capacity of the gel, mechanical strength of the gel, and production of the pigments by the entrapped cells, alginate and a photosensitive resin prepolymer (PVA-SbQ) were selected as the most suitable materials among the natural polysaccharides and synthetic resin prepolymers tested. *L. vera* cells entrapped in calcium alginate and the photosensitive resin grew well near the surface of the respective gels (Figs. 10 and 11). The respiratory activity which reached maximal values after an incubation of about 2 weeks was inferior to that of the free cells. Pigment production was observed by the entrapped cells under conditions similar to those for the free cells, the proliferation of the cells not being observed during the production of pigments. The maximal amount of the blue pigments was accumulated after an incubation of about 2 weeks in the presence of L-cysteine as an inducer. Although the photosensitive resin-entrapped *L. vera* cells produced the largest amount of pigments among the entrapped and free cells examined, in contrast to the calcium alginate-entrapped cells the majority of the pigments produced were retained in the gels. The pigment release from the photosensitive resin gels was enhanced to some extent by the increase in hydrophilicity of the gels with a higher saponification degree of poly-(vinyl alcohol), the main chain of the resin prepolymers (Table 9). Repeated use of the calcium alginate-entrapped cells for pigment production was successfully carried out for more than 7 months by the intermittent reactivation of the entrapped cells in the growth medium without L-cysteine. The entrapped cells prepared with the photosensitive resin were also able to produce repeatedly the blue pigments in several batches of incubation. The pigment productivity was improved by selecting the nitrogen source [124] and the carbon source. Under the optimized conditions, calcium alginate-entrapped *L. vera* cells could be applied to the continuous production of the pigments for at least 10 days in a jar bioreactor (Fig. 12) [125]. One of the precursors of the blue pigments, which turns blue by incubating with

Table 9. Effect of gel hydrophilicity on pigment production by photosensitive resin prepolymer (PVA-SbQ)-entrapped *Lavandula vera* cells [118]

Saponification degree of poly(vinyl alcohol) (%)	Total pigments produced (A_{590}/system)	Pigments in medium (%)
100	4.98—7.67	49.8—61.0
88	5.26—8.62	13.5—19.3
80	3.78—5.33	3.4— 5.8

Entrapped cells were incubated for 2 weeks in the medium (25 ml) containing 0.3 mM L-cysteine

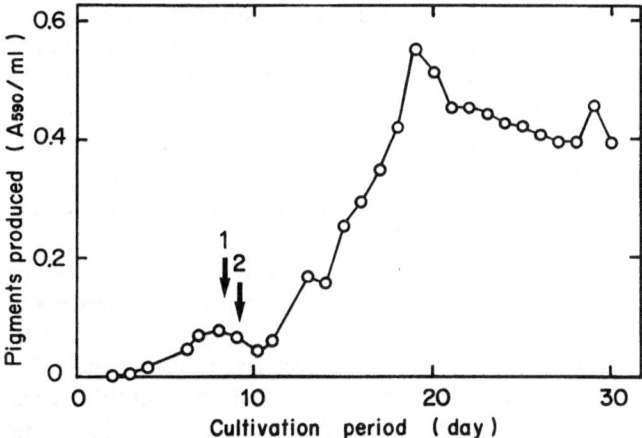

Fig. 12. Continuous production of blue pigments by entrapped *Lavandula vera* cells. Calcium alginate-entrapped cells were cultivated in a jar bioreactor (working volume, 1 l; aeration, 1 l min^{-1}). *Arrow 1*, addition of L-cysfeine (0.6 mM); *arrow 2*, start of continuous operation (flow rate, 250 ml day^{-1})

iron ions, was characterized as 1-methylethyl 3,4-dihydroxy-5-methoxy-3,10-diemethyl-4-oxo-4H-naphtho[2,3b]pyran-3-butanoate [126].

3.3 Immobilized Animal Cells

Animal cells have received increasing interest as useful biocatalysts for the production of vaccines and other bioactive peptides. However, application of animal cells to the production of useful substances has some drawbacks inherent to animal cells:

(1) Animal cells are fragile and sensitive to mechanical shock owing to a relatively large size of the cell and lack of a cell wall.

(2) The concentration of product per cell is regulated and maintained at a low level according to the mechanism of homeostasis. Furthermore, low cell concentrations (10^6 cells/ml) are obtained under *in-vitro* cultivation conditions due to the deficient supply of nutrients and accumulation of waste metabolites such as ammonia and lactate. Because of this low productivity and cell density the productivity by animal cells is usually quite low.

(3) Serum is essential to the cultivation of most animal cells as a growth factor although extensive efforts have been made to develop serum-free media. The supplementation of serum to the culture medium causes high costs for the cultivation of animal cells, and results in the contamination of the product with proteins derived from the serum and in difficulties in the purification of the product from the culture medium.

(4) Many types of animal cells except certain blood cells and hybridoma cells require an attachment to a surface for subsequent proliferation.

Although the development of serum-free media and improvement of culture systems have played important roles in overcoming these problems, immobilization of animal cells seems to be the most promising technique in order to solve some of the problems mentioned above. Namely, immobilization is able to provide a suitable surface for attachment of animal cells, to protect animal cells from external impacts, and to eliminate problems of productivity to some extent by repeated or continuous use of immobilized animal cells under optimized conditions as long as the product is excreted from the cells.

Production of mammalian bioactive peptides has recently become possible by using genetically manipulated microorganisms. Bioactive peptides synthesized in animal cells have complex structures, most of them being glycoproteins whose formation includes glycosylation of apoproteins. Correct glycosylation and other structural modifications cannot be, however, carried out in microbial cells. In some cases, apoproteins do not express sufficiently high activities in animal bodies and can be unstable. Utilization of animal cells as biocatalysts is, therefore, advantageous for the production of correctly modified proteins.

Immobilization of animal cells has been studied hitherto by using various methods, such as adsorption on ceramics and microcarriers, entrapment with gel-forming materials, encapsulation within membranes, and entrapment in hollow fibers [127]. Since many reviews have been published on the cultivation of anchor-dependent mammalian cells with microcarriers and the application of mammalian cells immobilized in hollow fibers and microcapsules was summarized by Heath and Belfor [128], this section mainly deals with the cells entrapped with polymer gel materials (Table 10).

Immobilized hybridoma LSP 21 cells and lymphoblastoid MLA 144 cells were prepared by entrapment in agarose beads, which were formed upon cooling of a cell/agarose suspension dispersed in paraffin oil. The immobilized hybridoma cells were able to produce monoclonal $IgG2A_k$ antibody against *Herpes simplex* type-2 virus glycoprotein for 7 days in a spinner vessel under daily replacement of about half of the medium. Continuous production of interleukin 2 was achieved for

Table 10. Application of immobilized mammalian cells

Mammalian cell	Gel material	Product	Ref.
Lymphoblastoid cell	Agarose	Interleukin 2	[129]
Rat pancreas islet	Crosslinked alginate	Insulin	[130]
Baby hamster kidney etc.	Calcium or strontium alginate	Erythropoietin	[131]
Mouse myeloma MPC-11	Reticulated polyvinyl formal resin	IgG	[132]
Hybridoma cell LSP21	Agarose	Monoclonal $IgG2A_k$	[129]
Hybridoma cell AcV_1	Calcium alginate	Monoclonal IgG_1	[133]
Hybridoma cell 4H11	Calcium alginate	Monoclonal IgA	[134]
Hybridoma cell 4C10B6	Calcium alginate	Monoclonal IgG	[135]
Hybridoma cell 16-3F	Calcium alginate + polyurethane	Anti-α-amylase monoclonal antibody	[136]

more than 13 days and the titer of the culture medium was maintained when the immobilized lymphoblastoid cells were cultivated in a spinner flask by replacing 80% of the medium every day [129].

Recently, several kinds of monoclonal antibodies have been reported to be synthesized by hybridoma cells entrapped with alginate gels [133, 134, 135]. By coating calcium alginate gels with urethane prepolymers, the immobilized cells could be cultivated in a fluidized-bed reactor for a long period of time [136].

Islets of the pancreas isolated from rats by the conventional collagenase digestion technique were enclosed in microcapsular membranes composed of cross-linked alginate. These membranes were prepared by treatment of the surfaces of the inner matrix of islet-containing calcium alginate beads with polylysine and poly-ethyleneimine. These entrapped islets were able to secrete insulin, whose release pattern responding to environmental glucose concentration was comparable to that of the control islets. The entrapped islets remained morphologically and functionally intact throughout a period of over 15 weeks. Furthermore, the glucose concentration in the blood of diabetic rats was maintained at a low level for almost 3 weeks when the entrapped islets were implanted into diabetic rats induced by streptozotocin. On the contrary, the free islets survived only for 6 to 8 days in diabetic rats [130]. Such studies are interesting for the future construction of bio-artificial endocrine organs by using immobilized animal cells.

Erythropoietin (EPO), a stimulator of erythrocyte production, was produced continuously for more than a month by calcium alginate-entrapped baby hamster kidney BHK21 cells, and by strontium alginate gel-entrapped NIH/3T3 cells which had previously been transformed with the EPO expression vector [131]. L929 cells from mouse connective tissues were also applicable to the production of EPO. To attain a high concentration of the cells and, subsequently, a high productivity of EPO, the formation of clear vacant spaces (channels) inside the alginate gels was demonstrated to be essential.

4 Recent Topics on Immobilized Growing Cells

Various types of living cells have been used to study their immobilization and their application to produce useful substances, as described in the previous sections. In particular, the immobilization of living cells of higher and multicellular organisms, such as plant and animal cells, has been exploited recently to overcome the fragility of these cells.

Immobilization of genetically improved microbial cells and their application have also been attempted. The recombinant DNA technique is an attractive method for the qualitative improvement of living cells, since this technique is able to provide a remarkable increase in the activities of certain target enzymes. It also provides novel catalytic functions, which have never been exhibited in the original organism, and it allows the potent synthesis of foreign peptides. At present, genetic manipulations are actively being carried out mostly on microbial cells. Consequently, genetically improved microbial cells are considered to be excellent biocatalysts for the production of useful substances. Hitherto, the application of such immobilized genetically improved microbial cells has not been attempted except for the few examples shown in Table 11.

Klein and Wegner [137] reported on the immobilization of genetically improved *Escherichia coli* cells with various gel materials and the application of the immobilized cells for the production of 6-aminopenicillanic acid from penicillin G. The penicillin G acylase activity of this *E. coli* strain, carrying the gene of penicillin G acylase, was 45 times higher than that of the non-induced and 8 times higher than that of the induced wild-type strain. Of the various immobilization methods tested, the epoxy bead preparation was found to be superior as judged from the cell-loading capacity, mechanical strength, and specific reaction rate of the immbilized cells.

Kimura [138] demonstrated the production of glutathione by *E. coli* carrying a plasmid, which consisted of plasmid pBR325 harboring the genes of two enzymes required for the biosynthesis of glutathione. *E. coli* cells entrapped with $\varkappa$-carrageenan could produce glutathione (6 g l^{-1}) at a 100% conversion in 20 mM L-cysteine in the presence of 80 mM L-glutamic acid. The conversion ratio of the immobilized cells was maintained at approximately 90% after 5 repeated uses.

E. coli BZ18 harboring a pTG201 plasmid derivative, which expressed catechol 2,3-dioxygenase activity, was immobilized in $\varkappa$-carrageenan beads and the immobilized cells were employed for the bioconversion of catechol to 2-hydroxymuconic semialdehyde in a packed bed reactor [139]. The results obtained revealed that the immobilized cells were more efficient biocatalysts than the free cells due to the difference observed between the cell density in suspension (8×10^8 cell ml^{-1}) and in gels (1.7×10^{11} cell ml^{-1}) after cell growth. Futhermore, the stability of the recombinant plasmid in *E. coli* cells in the absence of antibiotic selection was enhanced by immobilization of the host cells [147]. Stabilization of plasmid in cells by immobilization was also demonstrated when recombinant *E. coli* cells were cultivated in the absence of antibiotics for the production of amylase [140].

Table 11. Application of immobilized genetically improved microbial cells

Product	Microorganism	Gel material	Ref.
6-Amino penicillanic acid	*Escherichia coli*	Epoxide resin	[137]
Glutathione	*Escherichia coli*	$\varkappa$-Carrageenan	[138]
2-Hydroxymuconic semialdehyde	*Escherichia coli*	$\varkappa$-Carrageenan	[139]
Amylase	*Escherichia coli*	Dimethylsiloxane prepolymer	[140, 141]
β-Lactamase	*Escherichia coli*	Hollow fiber	[87]
		Calcium alginate	[142]
Polygalacturonate lyase, pH 2.5 acid phosphatase	*Myxococcus xanthus*	$\varkappa$-Carrageenan	[143]
Rat proinsulin	*Bacillus subtilis*	Agarose	[144]
Human proinsulin	*Escherichia coli*	Agarose	[145]
Somatomedin C	Yeast	Polyacrylamide-hydrazide	[146]

Continuous production of a recombinant protein (β-lactamase) was tried by using immobilized *E. coli* RB791 cells containing plasmid pKK, which is pBR322 modified by addition of the *tac* promotor upstream of the β-lactamase gene [142]. Production of extracellular foreign proteins (polygalacturonate lyase from *Erwinia chrysanthemi* and pH 2.5 acid phosphatase from *E. coli*) in *Myxococcus xanthus* immobilized growing cells was reported [143].

In the meantime, microbial cells carrying plasmid-encoding genes derived from animals were successfully applied for the extracellular production of bioactive compounds. That is, agarose-entrapped *Bacillus subtilis* cells carrying a plasmid harboring the rat preproinsulin gene and the penicillinase signal sequence of *Bacillus licheniformis* were able to synthesize and excrete proinsulin into the medium upon addition of the antibiotic novobiocin. Continuous formation of proinsulin (7–10 ng/ml) over a period of 80 h was observed at a flow rate of 4 ml/h under the prevention of cell growth by novobiocin [144]. Human proinsulin was produced in shake flask cultures by *E. coli* cells carrying the plasmid ptrc90K8 encoding the same polypeptide. The maximum production was about 2 mg l^{-1} day^{-1} [145].

Yeast cells harboring the plasmid p336/1, which contains the yeast α-mating factor gene and a synthetic somatomedin C gene, were able to produce and excrete continuously the growth hormone somatomedin C for over 10 days, when the cells were applied in an immobilized form in a flow-through reactor [146]. Erythropoietin production by genetically improved mammalian cells [131] was described previously.

Because of their high molecular weights, peptides and proteins produced by immobilized cells are likely to be retained inside the gels. Therefore, the selection of adequate gel materials to immobilize microbial and mammalian cells, genetically improved or not, is very important to construct efficient production systems.

5 Future Prospects

Since the old days, we have received great benefits from bioprocesses used in the preparation of foods, medicines, fuels, flavorings, coloring matters, etc. The introduction of bioprocesses into the chemical industry might lead to the substitution of conventional chemical synthetic processes, aiming at saving energy and resources, and the reduction of pollution. In response to these demands, a number of bioprocesses have been applied to the chemical, food, and pharmaceutical industries. At the same time, novel types of biocatalysts have been devised and developed. Immobilized growing cells are thought to be among the most promising biocatalysts for the application of bioprocesses, and many investigations have been reported on immobilized growing cells hitherto as mentioned above.

In the meantime, the genetic improvement of various cells has been actively studied using the techniques of gene manipulation and cell fusion. Immobilization techniques are also considered to have a great contribution to the application of these improved cells to modern biotechnology as described in this article. In future, various kinds of improved cells applicable to the desired bioprocesses should be screened, and these improved cells are expected to be used in the form of immobilized growing cells to construct more efficient catalytic systems. Futhermore, new types of methods or materials for the immobilization of different biocatalysts should be developed for the extensive application of the immobilized biocatalysts. Thus, intimate cooperation between qualitative improvement of biocatalysts by new techniques and development of immobilization techniques is essential for the application of biocatalysts to modern biotechnology in various fields.

6 References

1. Fukui S, Tanaka A (1982) Ann. Rev. Microbiol. 36: 145
2. Mosbach K (1987) (ed) Methods Enzymol. Vol. 135
3. Skryabin GK, Koshcheenko KA (1987) Methods Enzymol. 135: 198
4. Freeman A (1987) Methods Enzymol. 135: 216
5. Fukui S, Tanaka A, Iida T, Hasegawa E (1976) FEBS Lett. 66: 179
6. Fukushima S, Nagai T, Fujita K, Tanaka A, Fukui S (1978) Biotechnol. Bioeng. 20: 1465
7. Fukui S, Tanaka A (1984) Adv. Biochem. Eng./Biotechnol. 29: 1
8. Fukui S, Sonomoto K, Tanaka A (1987) Methods Enzymol. 135: 230
9. Ichimura K (1984) J. Polym. Sci. Polym. Chem. Edn. 22: 2817
10. Kumakura M, Kaetsu I, Nisizawa K (1984) Biotechnol. Bioeng. 26: 17
11. Slowinski W, Charm SE (1973) Biotechnol. Bioeng. 15: 973
12. Constantinides A, Bhatia D, Vieth WR (1981) Biotechnol. Bioeng. 23: 899
13. Wada M, Uchida T, Kato J, Chibata I (1980) Biotechnol. Bioeng. 22: 1175
14. Sarker JM, Mayaudon J (1983) Biotechnol. Lett. 5: 201
15. Fujimura M, Kato J, Tosa T, Chibata I (1984) Appl. Microbiol. Biotechnol. 19: 79
16. Tanaka T, Yamamoto K, Towprayoon S, Nakajima H, Sonomoto K, Yokozeki K, Kubota K, Tanaka A (1989) Appl. Microbiol. Biotechnol. 30: 564
17. Vaija J, Linko YY, Linko P (1982) Appl. Biochem. Biotechnol. 7: 51
18. Eikmeier H, Westmeier F, Rehm HJ (1984) Appl. Microbiol. Biotechnol. 19: 53
19. Horitsu H, Adachi S, Takahashi Y, Kawai K, Kawano Y (1985) Appl. Microbiol. Biotechnol. 22: 8
20. Lee YH, Lee CW, Chang NH (1989) Appl. Microbiol. Biotechnol. 30: 141
21. Horitsu H, Takahashi Y, Tsuda J, Kawai K, Kawano Y (1983) Eur. J. Appl. Microbiol. Biotechnol. 18: 358
22. Kautola H, Vahvaselka M, Linko YY, Linko P (1985) Biotechnol. Lett. 7: 167
23. Kautola H, Vassilev N, Linko YY (1989) Biotechnol. Lett. 11: 313
24. Stenroos SL, Linko YY, Linko P (1982) Biotechnol. Lett. 4: 159
25. Roy TBV, Blanch HW, Wilke CR (1982) Biotechnol. Lett. 4: 483
26. Boyaval P, Goulet J (1988) Enzyme Microb. Technol. 10: 725
27. Tipayang P, Kozaki M (1982) J. Ferment. Technol. 60: 595
28. Hang YD, Hamamci H, Woodams EE (1989) Biotechnol. Lett. 11: 119
29. Heinrich M, Rehm HJ (1982) Eur. J. Appl. Microbiol. Biotechnol. 15: 88
30. Venkatasubramanian K, Constantinides A, Vieth WR (1978) Enzyme Eng. 3: 29
31. Kennedy JF, Humphreys JD, Barker SA, Greenshields RN (1980) Enzyme Microb. Technol. 2: 209
32. Okuhara A (1985) J. Ferment. Technol. 63: 57
33. Ghommidh C, Navaro JM, Durand G (1981) Biotechnol. Lett. 3: 93
34. Ghommidh C, Navaro JM (1982) Biotechnol. Bioeng. 24: 1991
35. Kondo M, Suzuki Y, Kato H (1988) Hakko Kogaku Kaishi 66: 393
36. Ationu A, Patterson JDE, Todd JR, Wood BJB (1988) Biotechnol. Lett. 10: 671
37. Osuga J, Mori A, Kato J (1984) J. Ferment. Technol. 62: 139

38. Mori A (1985) Process Biochem. 20(3): 67
39. Nanba A, Kimura K, Nagai S (1985) J. Ferment. Technol. 63: 175
40. Brodelius P, Hagerdal B, Mosbach K (1980) Enzyme Eng. 5: 383
41. Szwajcer E, Brodelius P, Mosbach K (1982) Enzyme Microb. Technol. 4: 409
42. Wikström P, Szwajcer E, Brodelius P, Nilsson K, Mosbach K (1982) Biotechnol. Lett. 4: 153
43. Yi ZH, Rehm HJ (1982) Eur. J. Appl. Microbiol. Biotechnol. 16: 1
44. Morikawa Y, Karube I, Suzuki S (1979) Biotechnol. Bioeng. 21: 261
45. Deo YM, Gaucher GM (1984) Biotechnol. Bioeng. 26: 285
46. Kurzatkowski W, Kurylowicz W, Penyige A (1984) Appl. Microbiol. Biotechnol. 19: 312
47. Freeman A, Aharonowitz Y (1981) Biotechnol. Bioeng. 23: 2747
48. Khang YH, Shankar H, Senatore F (1988) Biotechnol. Lett. 10: 719
49. Khang YH, Shankar H, Senatore F (1988) Biotechnol. Lett. 10: 867
50. Morikawa Y, Karube I, Suzuki S (1980) Biotechnol. Bioeng. 22: 1015
51. Veelken M, Pape H (1982) Eur. J. Appl. Microbiol. Biotechnol. 15: 206
52. Foster BC, Coutts RT, Pasutto FM, Dossetor JB (1983) Biotechnol. Lett. 5: 693
53. Arcuri EJ, Nichols JR, Brix TS, Santamarina VG, Buckland BC, Drew SW (1983) Biotechnol. Bioeng. 25: 2399
54. Berk D, Behie LA, Jones A, Lesser BH, Gaucher M (1984) Can. J. Chem. Eng. 62: 112
55. Berk D, Behie LA, Jones A, Lesser BH, Gaucher M (1984) Can. J. Chem. Eng. 62: 120
56. Deo YM, Gaucher GM (1985) Appl. Microbiol. Biotechnol. 21: 220
57. Ogaki M, Sonomoto K, Nakajima H, Tanaka A (1986) Appl. Microbiol. Biotechnol. 24: 6
58. Takashima Y, Nakajima H, Sonomoto K, Tanaka A (1987) Appl. Microbiol. Biotechnol. 27: 106
59. Dalili M, Chau PC (1988) Biotechnol. Lett. 10: 331
60. Ohlson S, Flygare S, Larsson PO, Mosbach K (1980) Eur. J. Appl. Microbiol. Biotechnol. 10: 1
61. Sonomoto K. Hoq MM, Tanaka A, Fukui S (1981) J. Ferment. Technol. 59: 465
62. Sonomoto K, Hoq MM, Tanaka A, Fukui S (1983) Appl. Environ. Microbiol. 45: 436
63. Maddox IS, Dunnill P, Lilly MD (1981) Biotechnol. Bioeng. 23: 345
64. Sonomoto K, Nomura K, Tanaka A, Fukui S (1982) Eur. J. Appl. Microbiol. Biotechnol. 16: 57
65. Sonomoto K, Usui N, Tanaka A, Fukui S (1983) Eur. J. Appl. Microbiol. Biotechnol. 17: 203
66. Chun YY, Iida M, Iizuka H (1981) J. Gen. Appl. Microbiol. 27: 505
67. Kim JM, Sonomoto K, Tanaka A, Fukui S (1983) Ann. Rep. Internatl. Center Cooper. Res. Develop. Microbial Eng. Jpn. 6: 173
68. Larsson PO, Ohlson S, Mosbach K (1976) Nature 263: 796
69. Ohlson S, Larsson PO, Mosbach K (1978) Biotechnol. Bioeng. 20: 1267
70. Ohlson S, Larsson PO, Mosbach K (1979) Eur. J. Appl. Microbiol. Biotechnol. 7: 103
71. Kokubu T, Karube I, Suzuki S (1978) Eur. J. Appl. Microbiol. Biotechnol. 5: 233
72. Chevalier P, Noüe J (1987) Enzyme Microb. Technol. 9: 53
73. Shinmyo A, Kimura H, Okada H (1982) Eur. J. Appl. Microbiol. Biotechnol. 14: 7
74. Groom CA, Daugulis AJ, White BN (1988) Appl. Microbiol. Biotechnol. 28: 8
75. Chevalier P, de la Noüe J (1988) Enzyme Microb. Technol. 10: 19
76. Li GX, Linko YY, Linko P (1984) Biotechnol. Lett. 6: 645
77. Frein EM, Montenecourt BS, Eveleigh DE (1982) Biotechnol. Lett. 4: 287
78. Taniguchi M, Kato T, Matsuno R, Kamikubo T (1983) Eur. J. Appl. Microbiol. Biotechnol. 18: 218
79. Kumakura M, Tamada M, Kaetsu I (1984) Enzyme Microb. Technol. 6: 411
80. Jirků V, Vojtisek V, Veruovic B, Kubanek V, Kralicek J (1984) Biotechnol. Lett. 6: 363
81. McHale AP (1988) Biotechnol. Lett. 10: 361
82. Kokubu T, Karube I, Suzuki S (1981) Biotechnol. Bioeng. 23: 29
83. Vuillemard JC, Terre S, Benoit S, Amiot J (1988) Appl. Microbiol. Biotechnol. 27: 423
84. Aleksieva P, Tchorobanov B, Suchodol'skaya G, Koschteenko K (1988) Appl. Microbiol. Biotechnol. 29: 239

85. Younes G, Nicaud JM, Guespin-Michel J (1984) Appl. Microbiol. Biotechnol. 19: 67
86. Okada T, Sonomoto K, Tanaka A (1987) Biochem. Biophys. Res. Commun. 145: 316
87. Inloes DS, Smith WJ, Taylor DP, Cohen SN, Michaels AS, Robertson CR (1983) Biotechnol. Bioeng. 25: 2653
88. Linko S (1988) Enzyme Microb. Technol. 10: 410
89. Yongsmith B, Sonomoto K, Tanaka A, Fukui S (1982) Eur. J. Appl. Microbiol. Biotechnol. 16: 70
90. Okada T, Sonomoto K, Tanaka A (1987) Appl. Microbiol. Biotechnol. 26: 112
91. Nojima S (1983) Chem. Econ. Eng. Rev. 15(4): 17
92. Sakamoto M, Iida T, Izumida H, Takiguchi K (1988) Proc. 8th Internatl. Symp. Alcohol Fuels, p. 15
93. Matsui S, Sejima S, Izumida H (1988) Proc. 8th Internatl. Symp. Alcohol Fuels, p. 107
94. Krouwel PG, Groot WJ, Kossen NWF, Laan WNM (1983) Enzyme Microb. Technol. 5: 46
95. Karube I, Matsunaga T, Tsuru S, Suzuki S (1976) Biochim. Biophys. Acta 444: 338
96. Karube I, Kuriyama S, Matsunaga T, Suzuki S (1980) Biotechnol. Bioeng. 22: 847
97. Kumar PKR, Lonsane BK (1988) Appl. Microbiol. Biotechnol. 28: 537
98. Sauced JEN, Barbotin JN, Thomas D (1989) Appl. Microbiol. Biotechnol. 30: 226
99. Brodelius P, Deus B, Mosbach K, Zenk MH (1979) FEBS Lett. 103: 93
100. Alfermann AW, Schuller I, Reinhardt E (1980) Planta Medica 40: 218
101. Jones A, Veliky IA (1981) Eur. J. Appl. Microbiol. Biotechnol. 13: 84
102. Veliky IA, Jones A (1981) Biotechnol. Lett. 3: 551
103. Felix H, Brodelius P, Mosbach K (1981) Anal. Biochem. 116: 462
104. Galun E, Aviv D, Dantes A, Freeman A (1983) Planta Medica 49: 9
105. Wichers HJ, Malingre TM, Huizing HJ (1983) Planta 158: 482
106. Wichers HJ, Malingre TM, Huizing HJ (1985) Planta 165: 264
107. Furuya T, Yoshikawa T, Taira M (1984) Phytochem. 23: 999
108. Brodelius P, Nilsson K (1983) Eur. J. Appl. Microbiol. Biotechnol. 17: 275
109. Majerus F, Pareilleux A (1986) Biotechnol. Lett. 8: 863
110. Brodelius P, Nilsson K (1980) FEBS Lett. 122: 312
111. Shuler ML (1981) Ann. N. Y. Acad. Sci. 369: 65
112. Jose W, Pedersen H, Chin CK (1983) Ann. N. Y. Acad. Sci. 413: 409
113. Jirků V, Macek T, Vanek T, Krumphanzl V, Kubanek V (1981) Biotechnol. Lett. 3: 447
114. Haldimann D, Brodelius P (1987) Phytochem. 26: 1431
115. Lindsey K, Yeoman MM, Black GM, Mavituna F (1983) FEBS Lett. 155: 143
116. Barnabas NJ, David SB (1988) Biotechnol. Lett. 10: 593
117. Nakajima H, Sonomoto K, Usui N, Sato F, Yamada Y, Tanaka A, Fukui S (1985) J. Biotechnol. 2: 107
118. Nakajima H, Sonomoto K, Sato F, Morikawa H, Ichimura K, Yamada Y, Tanaka A (1986) Appl. Microbiol. Biotechnol. 24: 266
119. Ayabe S, Iida K, Furuya T (1986) Plant Cell Rep. 3: 186
120. Miyasaka H, Nasu M, Yamamoto T, Endo Y, Yoneda K (1986) Phytochem. 25: 1621
121. Kobayashi Y, Fukui H, Tabata M (1987) Plant Cell Rep. 6: 185
122. Kobayashi Y, Fukui H, Tabata M (1988) Plant Cell Rep. 7: 249
123. Watanabe K, Sato F, Furuta M, Yamada Y (1985) Agric. Biol. Chem. 49: 533
124. Nakajima H, Sonomoto K, Sato F, Ichimura K, Yamada Y, Tanaka A (1989) J. Ferment. Bioeng. 67: 306
125. Nakajima H, Sonomoto K, Sato F, Yamada Y, Tanaka A (1989) Agric. Biol. Chem. 53: 3077
126. Nakajima H, Sonomoto K, Sato F, Yamada Y, Tanaka A (1990) Agric. Biol. Chem. 54: 53
127. Nilsson K (1987) Trends Biotechnol. 5: 73
128. Heath C, Belfort G (1987) Adv. Biochem. Eng./Biotechnol. 34: 1
129. Nilsson K, Scheirer W, Merten OW, Ostberg L, Liehl E, Katinger HWD, Mosbach K (1983) Nature 302: 629
130. Lim F, Sun AM (1980) Science 210: 908

131. Shirai Y, Sasaki R, Hashimoto K, Kawahara H, Hitomi K, Chiba H (1988) Appl. Microbiol. Biotechnol. 29: 544
132. Yamaji H, Fukuda H, Nojima Y, Webb C (1989) Appl. Microbiol. Biotechnol. 30: 609
133. Bugarski B, King GA, Daugulis AJ, Goosen MFA (1989) Appl. Microbiol. Biotechnol. 30: 264
134. Shirai Y, Hashimoto K, Yamaji H, Tokashiki M (1987) Appl. Microbiol. Biotechnol. 26: 495
135. Shirai Y, Hashimoto K, Yamaji H, Kawahara H (1988) Appl. Microbiol. Biotechnol. 29: 113
136. Iijima S, Mano T, Taniguchi M, Kobayashi T (1988) Appl. Microbiol. Biotechnol. 28: 572
137. Klein J, Wagner F (1980) Enzyme Eng. 5: 335
138. Kimura A (1984) Kagaku Zokan 103: 123
139. Dhulster P, Barbotin JN, Thomas D (1984) Appl. Microbiol. Biotechnol. 20: 87
140. Oriel P (1988) Enzyme Microb. Technol. 10: 518
141. Oriel P (1988) Biotechnol. Lett. 10: 113
142. Georgiou G, Chalmers TJ, Shuler ML, Wilson DB (1985) Biotechnol. Progress 1: 75
143. Younes G, Breton AM, Guespin-Michel J (1987) Appl. Microbiol. Biotechnol. 25: 507
144. Mosbach K, Birnbaum S, Hardy K, Davis J, Bülow L (1983) Nature 302: 543
145. Birnbaum S, Bülow L (1988) Enzyme Microb. Technol. 10: 601
146. Sode K, Brodelius P, Meussdoerffer F, Mosbach K, Ernst JF (1988) Appl. Microbiol. Biotechnol. 28: 215
147. Poet PT, Dhulster P, Barbotin JN, Thomas D (1986) J. Bacteriol. 165: 871

New Developments in the Chemo-Enzymatic Production of Amino Acids

J. Kamphuis, W. H. J. Boesten, Q. B. Broxterman, H. F. M. Hermes, J. A. M. van Balken, E. M. Meijer, and H. E. Schoemaker
DSM Research, Bio-organic Chemistry Section, P.O. Box 18, 6160 MD Geleen, The Netherlands

Dedicated to Professor Armin Fiechter
on the occasion of his 65th birthday

Recent progress in the chemo-enzymatic production of amino acids is reviewed. Both recently developed commercial processes and potentially important new developments are discussed. Emphasis is placed on the use of acylases, aminopeptidases and hydantoinases. The discovery of D-specific enzymes in combination with racemases is an exciting and promising new area. Also, a goal-orientated approach towards the selective generation of these novel enzyme activities using in vivo protein engineering techniques is highlighted. The interest in dipeptide sweeteners has triggered a major research effort towards the production of L-phenylalanine and D-alanine. A number of methods for the production of these amino acids is briefly discussed. Finally, chemo-enzymatic methods for the synthesis of enantiomerically pure α-alkyl-α-amino acids are reviewed.

Advances in Biochemical Engineering/
Biotechnology, Vol. 42
Managing Editor: A. Fiechter
© Springer-Verlag Berlin Heidelberg 1990

1 General Introduction

From time immemorial, amino acids have played an important role in biology, biochemistry and (industrial) chemistry. Amino acids are the building blocks of proteins and they play an essential role in the regulation of the metabolism of living organisms.

The interest in the commercial production of amino acids dates back to 1908 with the discovery of monosodium glutamate (MSG) as a seasoning agent by Kikunae Ikeda (see [1]). At first MSG was industrially obtained from wheat, soybean and other plant proteins. Since 1957, however, microbial production processes have been successfully developed for glutamic acid and subsequently also for other proteinogenic amino acids like lysine, valine, proline, threonine, phenylalanine etc. However, a detailed discussion of these microbial processes and the application of novel genetic engineering techniques for strain improvement would go beyond the scope of the present review. In general, from a commercial viewpoint it can be stated that when comparing chemo-enzymatic methods with microbial processes, the latter are highly competitive for the large-scale production (thousands of tons per year) of naturally occurring proteinogenic L-amino acids, whilst for the production of natural amino acids on a small scale or for the production of synthetic D- and/or L-amino acids, chemo-enzymatic methods are the methods of choice in addition to all-chemical procedures.

Current interest in developing peptide-derived chemotherapeutics has heightened the importance of rare and non-proteinogenic enantiomerically pure amino acids. For example, D-phenylglycine and D-p-hydroxy-phenylglycine are building blocks for the broad spectrum β-lactam antibiotics ampicillin and amoxycillin, respectively. The natural amino acid L-valine is used as feedstock in the microbial production of the cyclic undecapeptide cyclosporin A, which has immuno-suppressive activity and is used in human transplant surgery.

Also, amino acids are versatile chiral building blocks for a whole range of fine chemicals. In the last two decades there has been a growing public awareness and concern with regard to the exposure of man and the environment to an ever increasing number of chemicals. The benefits, however, arising from the use of therapeutic agents, pesticides, food and feed additives, etc. are enormous. Hence there is still an ever increasing demand for more selective drugs and pesticides which are targeted in their mode of action, exhibit less toxic side-effects and are more environmentally acceptable [2, 3]. To this end, a central role will be played by chiral (optically pure) compounds since nature at the molecular level is intrinsically chiral. Consequently, this provides an important stimulus to companies to market chiral products as optical pure isomers. This in turn results in an increasing need for efficient methods for the industrial synthesis of optically active compounds. This is exemplified by the production of D-valine, an intermediate in the production of the pyrethroid insecticide Fluvalinate. Another example is L-homophenylalanine, which is a constituent of a number of recently developed Angiotensin-Converting-Enzyme (ACE) inhibitors.

The (bio)-synthesis and/or microbial production of amino acids has been the subject of a large number of monographs and review papers [e.g. 1, 4, 5]. The aim of the present review is to highlight some new developments in the chemo-enzymatic production of amino acids.

At first enzymatic kinetic resolution processes will be discussed. In addition and in part complementary to established processes based on the use of amino-acylases and hydantoinases, a novel and recently commercialized kinetic resolution process based on the stereospecific hydrolysis of amino acid amides by an aminopeptidase from *Pseudomonas putida* will be discussed. In addition, the review will highlight new developments in the improvement of established processes. The use of membrane reactors for the acylase-process and the use of a carbamoylase in the hydantoinase process, have been reviewed and will be mentioned briefly. Emphasis will be placed on the development of new enzyme systems such as D-specific acylases, L-specific hydantoinases and D-specific aminopeptidases. Both screening methods and in vivo protein engineering techniques will be discussed.

For an economic resolution process, efficient procedures for recycling of the unwanted isomer are mandatory. In addition to chemical methods and the well-known amino acid racemases, *N*-acylamino acid racemases and amino acid amide racemases have recently been described. Incorporation of those enzyme systems in respectively the acylase process or the aminopeptidase process allow for the essentially quantitative conversion of racemic starting material into the desired enantiomerically pure amino acid in a single operation.

The introduction of aspartame, or L-*N*-aspartyl-L-phenylalanine methyl ester, as a novel non-nutritional dipeptide sweetener has generated wide-spread interest both in the development of new dipeptide sweeteners and in the development of various chemo-enzymatic methods for the production of the constituent amino acids. As examples of the potential of chemo-enzymatic approaces for the synthesis of this type of compounds, various routes towards L-phenylalanine, the major constituent of aspartame, and D-alanine, constituent of a novel type of dipeptide sweetener, 'alitame', will be discussed.

In the last part of the review the chemo-enzymatic synthesis of the non-proteinogenic α-alkyl substituted amino acids will be discussed. In recent years medicinal chemists have been increasingly interested in bio-active peptide-analogs containing α-alkyl amino acids (e.g. neuropeptides, chemotactic peptides, enkephalins), mainly because they tend to freeze specific conformations and/or dramatically slow down enzymatic degradation processes. In this review novel chemo-enzymatic production methods will be discussed, one of which is based on the stereoselective hydrolysis of α-alkyl substituted amino acid amides with an aminopeptidase from *Mycobacterium neoaurum*.

2 Enzymatic Resolution Processes for the Production of α-H-Amino Acids

2.1 Applications of Amino Acids

Amino acids are versatile chiral building blocks used in the synthesis of pharmaceuticals, agro chemicals and food/feed additives. Illustrative examples are given in Figs. 1 and 2.

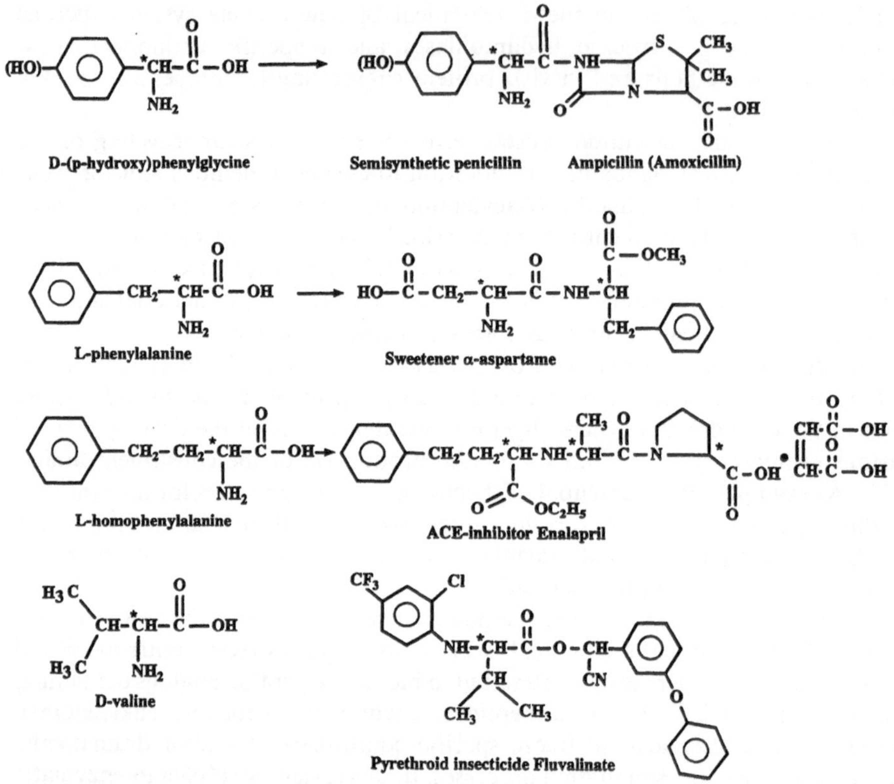

Fig. 1. Some illustrative examples of the applications of L- and D-α-H-amino acids

2.2 Acylases

Several L-amino acids are produced on a large scale by enzymatic resolution of N-acetyl-D,L-amino acids [6–11].

Acylase immobilized on DEAE-Sephadex is for example employed in a continuous process while Degussa uses the free acylase retained in a membrane-reactor [12, 13]. In the latter process the advantages of reuse of the enzyme and homogeneous catalysis are combined. However, the products have to be separated using ion

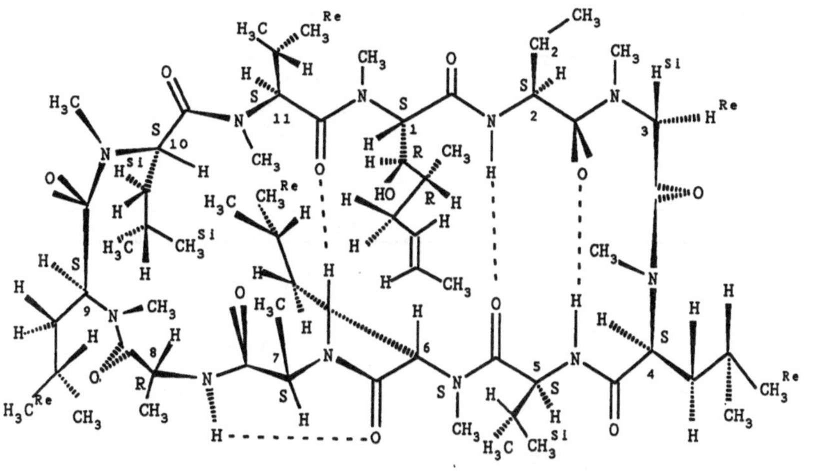

Fig. 2. The immunosuppressant Cyclosporin A

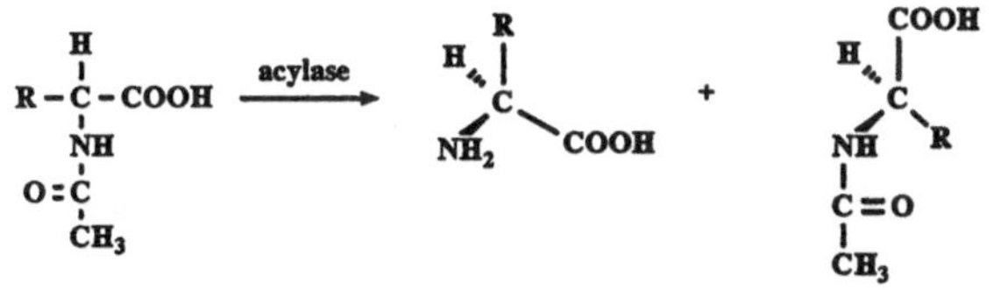

D,L-N-acetyl amino acid   L-amino acid   D-N-acetyl amino acid

Fig. 3. Commercial process for the enzymatic production of L- and D-amino acids from N-acetyl-D,L-amino acids

exchange columns and the starting material is a derivative rather than a precursor of the racemic amino acid, thus making the total process circuitous since it involves several chemical steps in addition to the enzymatic resolution step. Furthermore external racemization of the unwanted isomer is not easily accomplished. Also some of the N-acylamino acids are unsusceptible to the enzyme [14, 15].

Whitesides et al. [15] studied in detail the substrate specificity and enzyme characteristics of Acylase I (aminoacylase; N-acylamino-acid amidohydrolyse, EC 3.5.1.14.) from porcine kidney (PKA) and from an *Aspergillus* species (AA). These authors demonstrated that acylase I is a broadly applicable enzymatic catalyst for the kinetic resolution of unnatural and rarely occuring α-amino acids. This is nicely illustrated in Table 1.

The enantioselectivity of the acylase I is high (>90% enantiomeric excess (e.e., which is defined as the ratio R — S/R + S times 100%), but the system has several restrictions; if the hydrogen atom on the amide nitrogen is replaced by an alkyl group the substrate activity is destroyed; N-methyl and N-ethyl-N-acylamino

Table 1. Reactivities of Substituents in Substrates of Acylase I (Ref. 15)

$$R_2{}_{\prime\prime\prime}\ \underset{R_1}{\overset{CO_2H}{C}}\ \underset{H}{\overset{}{N}}\ \overset{O}{C}\text{—}R_3$$

R_1	R_2	R_3	reactivity[a]
CH₃(CH₂)₀₋₅, (CH₃)₂CH, ⟩CH₂, ⟩CH, ═(CH₂)₁₋₃, CH₃···CH₂, ⟩CH₂, ⟩(CH₂)₁₋₃, △(CH₂)₀₋₁, HOCH₂, (OH)···CH₂, ClCH₂, NC(CH₂)₃₋₄, HOOC(CH₂)₂₋₃, CH₃(CH₂)₀₋₁, S⟩CH₂, ⟨benzene⟩S(CH₂)₁₋₃, ⟨benzene⟩S(CH₂)₀₋₃, HO-⟨benzene⟩CH₂, ⟨acetyl⟩CH₂, ⟨furan⟩CH₂, ⟨indole⟩CH₂	H	CH₃, C₂H₅; XCH₂ (X=Cl, Br, CH₃O); XCH₂CH₂ (X=Cl, Br); H,[b] C₆H₅,[d]; H₂NCH₂[d]	good, >10 %
CH₃(CH₂)₆, ⟩(CH₂)₆, ⟨cyclohexenyl⟩CH,[a] HO-CH₃CH, H₂N-C(=O)-CH₂, HO(CH₂)₃₋₈	H	CH₃(CH₂)₀₋₁CH(Cl)[g]; L-RCH(NH₂)[g]	fair, 1-10 %
CH₃(CH₂)₇₋₈,[a] (CH₃)₃C, ⟨cyclohexyl⟩(CH₂)₁₋₂,[a] HO-C(CH₃)₂CH₂, ⟨imidazole⟩CH₂,[a] H₂N-C(NH₂⁺)-NH(CH₂)₃,[a] Cl-CH₂-C(=O)-NHCH₂,[a] H₃N⁺(CH₂)₃₋₄[a]	CH₃[b]	CH₃CH₂CH(CH₃)[b]	poor, 0.01-1 %

a: Data for PKA only
b: PKA only; no reactivity with AA

acids (tertiary amides) are not substrates, likewise, upon replacing amide nitrogen by oxygen no enzyme activity is observed; acetyllactic acid and mandelic acid, ester analogs of acyl amino acids, are not substrates. Other compounds not accepted as substrates include acyl derivatives of proline and aspartic acid.

The effect of the type of acyl group on the rate of enzymatic hydrolysis was also investigated. The results showed that enzymes from both animal (PKA) and microbial sources (AA) hydrolyze formyl, benzoyl and glycyl amino acids, although AA does so at relatively higher rates than those observed for PKA. PKA but not AA tolerates branching at the α-position of acylgroups. PKA therefore hydrolyzes dipeptides and α-chloroacetyl or α-methylacyl amino acids, whereas

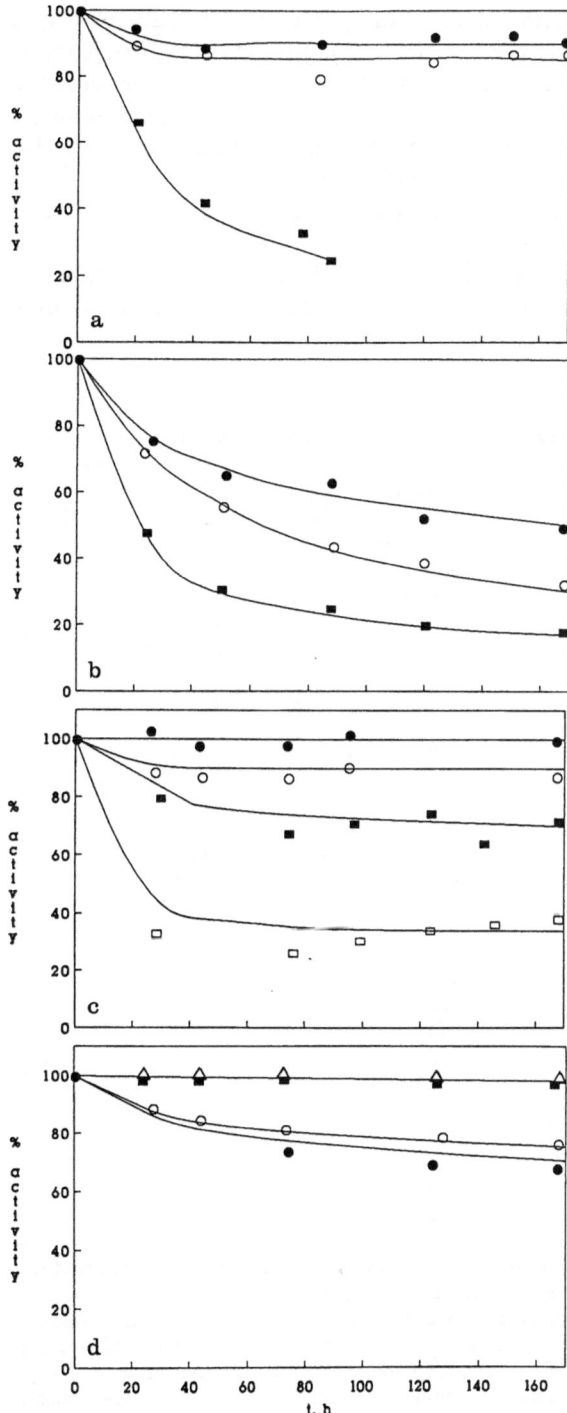

Fig. 4a–d. Stability of soluble acyclase I in the presence of organic cosolvents. Acyclase I from porcine kidney (**a, c**) and from *Aspergillus* (**b, d**) were incubated in 0.10 M phosphate buffer, pH 7.5, containing ethanol (a, b) or DMSO (c, d) at 25 °C, under nitrogen. Concentrations of organic cosolvents were 10% (●), 20% (○), 30% (■), 40% (□) and 50% (△) (Ref. 15)

AA does not (see Table 1). Furthermore it was proven that Cobalt (Co^{2+}), in general, activates AA, increasing the rates of reaction by 15–200%. PKA shows this effect also but less consistently. PKA and AA are zinc metalloproteins and are generally unaffected or inhibited by other divalent metal cations in solution. The mechanism of Co^{2+}-activation is unknown, but it appears not to involve simple substitution of Co^{2+} for Zn^{2+} in the enzyme active sites [16, 17].

In conclusion, Acylase I is a useful catalyst for the resolution of unnatural amino acids on a multigram scale.

Its stability in solution allows it to be used in soluble form, and membrane-enclosure of the enzyme allows large quantities of protein to be used to catalyze

Table 2. Substrate specifity of *Bacillus thermoglucosidius* amino acylase (Ref. 18)

Substrate	Relative activity[a]	K_m (mM)	V_{max} (μmol min^{-1} mg^{-1})
N-Acetyl-L-methionine	100	7.9	3.410
N-Acetyl-L-alanine	78	3.8	3.050
N-Acetyl-D,L-norleucine	70	6,4	2.170
N-Acetyl-L-valine	69	3.2	2.760
N-Acetyl-D,L-serine	28		
N-Acetyl-L-glutamine	27	1.1	740
N-Acetylglycine	18		
N-Acetyl-L-tyrosine	14		
N-Acetyl-L-leucine	12		
N-Acetyl-L-phenylalanine	3		
N-Chloroacetylglycine	640	22	35.200
N-Chloroacetyl-D,L-norleucine	640	23	34.600
N-Chloroacetyl-D,L-valine	620	28	30.600
N-Chloroacetyl-D,L-alanine	570		
N-Chloroacetyl-D,L-serine	440		
N-Chloroacetyl-L-leucine	370	9.7	10.200
N-Chloroacetyl-L-tyrosine	230		
N-Chloroacetyl-D,L-phenylalanine	52		
L-Alanyl-L-alanine	28		
L-Alanyl-L-leucine	9		
L-Alanylglycine	5		
Glycyl-L-leucine	5		
L-Valyl-L-alanine	3		
L-Leucyl-L-alanine	3		
L-Alanyl-L-alanyl-L-alanine	4		
L-Leucylglycylglycine	2		

[a] At the substrate concentration of 40 mM (for compounds with L-configuration) or 80 mM (for racemic compounds). Inert (relative activity of less than 1%):
N-Acetyl-L-aspartate, *N*-acetyl-L-glutamate, α-*N*-acetyl-L-lysine, ε-*N*-acetyl-L-lysine, *N*-acetyl-D-methionine, *N*-acetyl-D-leucine, *N*-acetyl-D-valine, *N*-*t*-butoxycarbonyl-L-methionine, *N*-benzyloxycarbonyl-L-methionine, L-histidyl-L-leucine, L-methionylglycine, L-leucylglycine, L-phenylalanyl-L-leucine, L-alanyl-L-alanyl-L-alanyl-L-alanine, glutathione, *N*-acetylglycyl-L-Leucine, *N*-benzyloxycarbonylglycyl-L-phenylalanine (15 mM), D,L-phenylalanine amide (25 mM), L-alanyl-L-alanine amide (25 mM) and L-leucine-*p*-nitroanilide (1 mM)

the reactions of even very low activity substrates. The practical laboratory application of PKA extends to the resolution of substrates having > 1 % of the activity
of acetyl-methionine on a 10–100 g scale and to the resolution of substrates having
0.001–1 % activity on a 1–5 g scale. The enzyme is remarkable stable in high
concentrations of organic solvents such as ethanol (up to 50%) and isopropanol
(up to 30%). Concentrations lower than 30% of ethanol activate the enzyme somewhat (see Fig. 4).

AA exhibits a somewhat more limited substrate specificity than PKA. In addition
to non-specific hydrolysis, a problem of this system is the incomplete hydrolysis
of the L-enantiomer, causing enantiomeric contamination of the D-product.
Substrates having high Km values (> 10 mM) will not afford highly enantiomerically enriched D-products without recrystallization. In general, though, these
resolutions provide, in a single process, both enantiomers of an amino acid having
high enantiomeric purities. Like any resolution, a disadvantage of the method is
that the maximum chemical yield of either enantiomer is 50%.

Soda et al. [18] reported that through screening in thermophilic bacteria a
thermostable L-amino acylase from *Bacillus thermoglycosidius* (DSM 2542) was
found. The purified enzyme has a molecular weight of about 175 000 and is composed of four subunits identical in molecular weight (43 000). The enzyme contains 4 g atoms of zinc per mole of enzyme protein. The enzyme catalyzes hydrolysis of various kinds of *N*-acyl-L-amino acids (see Table 2) with very high molecular
activity compared to those of fungal and mammalian enzymes. V_{max} and K_m for
N-acetyl-L-methionine are 3410 units mg^{-1} protein and 7.9 mM, respectively.

The enzyme lost no activity when incubated at 37 °C between pH 6.0 and 11.5
for 1 h. The enzyme retained its full activity on heating at 70 °C for 10 min and
90% of the original activity after heating at 80 °C (see Fig. 5).

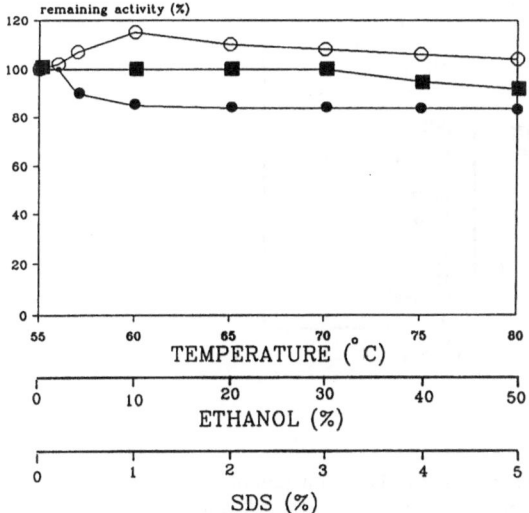

Fig. 5. Stability of the thermostable
acyclase from *Bacillus thermoglycosidius* (■) incubation at various temperatures for 10 min, (O) at 37 °C for
20 min in the presence of various
concentrations of ethanol and SDS
(●) (Ref. 18)

2.3 Amidases/Aminopeptidases

Maestracci et al. [19] reviewed the amidases from *Brevibacterium* strains. However, no mention was made of the very efficient and universally applicable industrial process developed by DSM [20–24], which is used for the production of both optically pure L- and D-amino acids. Pivotal in this process is the enantioselective hydrolysis of D,L-amino acid amides. The stable D,L-amino acid amides are efficiently prepared under mild reaction conditions starting from simple raw materials (see Fig. 6).

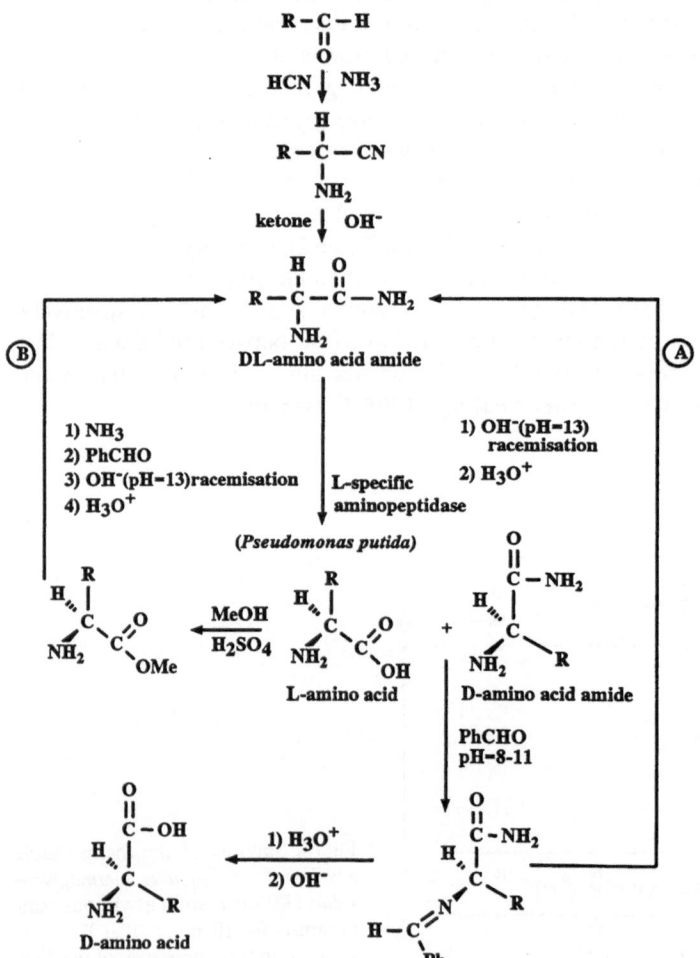

Fig. 6. Reaction scheme for DSM industrial process for the production of a range of L- as well as D-α-amino acids. *A* and *B* are the recycle procedures for the unwanted isomer

The reaction of an aldehyde with hydrogen cyanide in ammonia (Strecker reaction) affords the amino nitrile. The amino nitrile is converted in a high yield to the D,L-amino acid amide under alkaline conditions in the presence of a catalytic amount of acetone [25]. The resolution step [26] is accomplished with permeabilized whole cells of *Pseudomonas putida* ATCC 12633 and a nearly 100% stereoselectivity in hydrolyzing only the L-amino acid amide is combined with a very broad substrate specificity (see Table 3).

Not only the smallest optically active amino acid i.e. alanine, but also valine, leucine, several (substituted) aromatic amino acids, heterosubstituted amino acids [methionine, homomethionine [27] and thienylglycine] and even an imino acid, proline are obtainable in both the L- and D-forms [28]. No enzymic side effects are observed and substrate concentrations up to 20% by weight can be used without affecting the enzyme activity. The biocatalyst is active in a broad pH-range and can be used in soluble form in a batchwise process, thus poorly soluble amino acids can be resolved without technical difficulties. Re-use of the biocatalyst is in principle possible. A very simple and elegant alternative to the use of ion-exchange columns or extraction to separate the mixture of D-amino acid amide and the L-amino acid has been elaborated at DSM. Thus addition of one equivalent of benzaldehyde (with respect to the D-amino acid amide) to the enzymic hydrolysate results in the formation of a Schiff base with the D-amino acid amide which is insoluble in water and, therefore, can be easily separated [29]. Acid hydrolysis (H_2SO_4, HX, HNO_3 etc.) results in the formation of the D-amino acid (without racemization). Alternatively the D-amino acid amide can be hydrolysed by cell-preparations of *Rhodococcus erythropolis* [30]. This

Table 3. Substrate specificity of the aminopeptidase from *Pseudomonas putida* ATCC 12633 (selected examples)

biocatalyst lacks stereoselectivity. This option is very useful for amino acids which are highly soluble in the neutralized reaction mixture obtained after acid hydrolysis of the amide. Process economics dictate the recycling of the unwanted isomer. Path A in Fig. 6 illustrates that racemization of the D-*N*-benzylidene amino acid amide is facile and can be carried out under very mild reaction conditions [31]. After removal of the benzaldehyde the D,L-amino acid amide can be recycled. This option shows that 100% conversion to the L-amino acid is theoretically possible. A suitable method for racemisation and recycling of the L-amino acid (path B, Fig. 6) comprises the conversion of the L-amino acid into the ester in the presence of concentrated acid, followed by addition of ammonia, resulting in the formation of the amide. Addition of benzaldehyde and racemisation by base (pH = 13) gives the D,L-amino acid amide. In this way 100% conversion to the D-amino acid is possible. The presence of an α-hydrogen atom in the substrate is an essential structural feature for the enzymic activity of the aminopeptidase of *Pseudomonas putida*. Its enantioselectivity for α-H-amino acid amides is absolute, yet it accepts substrates having a wide range of structures and functionalities. This is illustrated in Table 4 where the results of some recently performed kinetic resolutions are shown.

L- and D-lupinic acid have been prepared for the first time in collaboration with van der Plas et al. [32]. The β-purinyl-L-alanine derivative is isolated from *Lupinus angustifolins* [33] which is a principle metabolite of the phytohormone transzealine (E-16), being one of the most effective natural stimulants of plant cell division.

The distinct advantages of the aminopeptidase process are the following, the substrate for the enzymatic hydrolysis is a precursor of the amino acid, the number of chemical steps can be kept to a minimum, the use of relatively cheap whole cell biocatalyst contributes to the economical feasibility of the procedure, and L- as well as D-amino acids can be prepared with a very high optical purity. Moreover, the enantiomerically pure Schiff bases of D-amino acid amides are versatile chiral building blocks. In collaboration with Ottenheijm et al. the syn-

Table 4. Some examples of enantiomerically pure trifunctional amino acids which have been obtained in the L- as well as D-form using the aminopeptidase technology of DSM

thesis of chiral *N*-hydroxyamino acid amides from the Schiff bases has been studied [34].

N-Hydroxy α-amino acids derivatives are widely found in nature [35]. They can be found, among others, as constituents of peptides, to which physiological properties can be attributed such as antibiotic activity [36]. Two oxidative preparation methods for optically active *N*-hydroxy-α-amino acid amides have been developed [34].

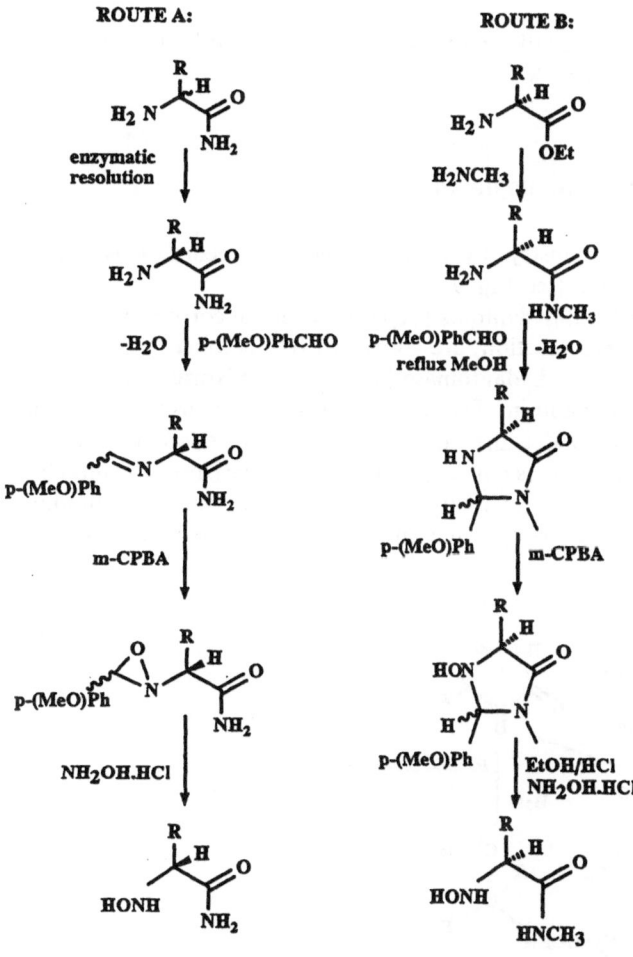

ROUTE A: **ROUTE B:**

R-i-Pr, i-But, Ph, CH₂Ph, CH₂CH₂Ph   R-Me, i-But, Ph, CH₂Ph

Fig. 7. Two oxidative routes for the synthesis of enantiomerically pure *N*-hydroxy amino acid amides

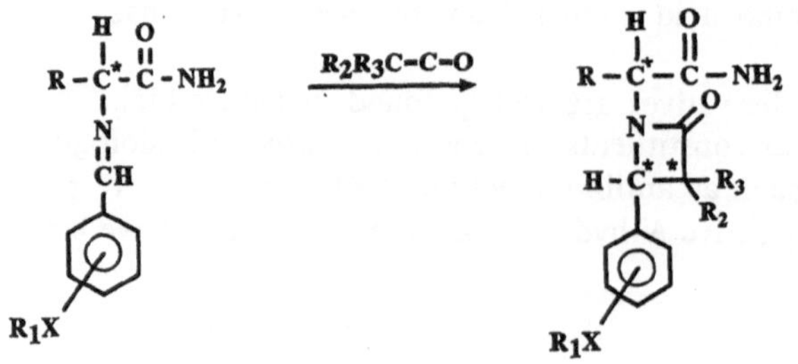

Fig. 8. Cycloaddition reactions of chiral Schiff bases

The Schiff's base are also versatile synthetic tools for the synthesis of β-lactam derivatives as illustrated in Fig. 8.

2.4 Hydantoinases/Decarbamoylases

Another method used in industry, by Kanegafuchi and Recordati, is the application of D-hydantoinases (see Fig. 9).

Chemically synthesized D,L-hydantoins prepared from the corresponding aldehydes via the Bucherer-Berg reaction, are converted by microbial cells (*Bacillus brevis*), containing a D-specific hydantoinase [37–42] to a mixture of D-*N*-carbamoyl amino acid and L-hydantoin. The latter compound undergoes rapid and spontaneous racemization under the conditions of the reaction, therefore in principle 100% of the hydantoin is converted into the D-*N*-carbamoyl compound. The D-amino acid is obtained after treatment of the D-*N*-carbamoyl compound with nitrous acid. This process is operated on an industrial scale by the Japanese

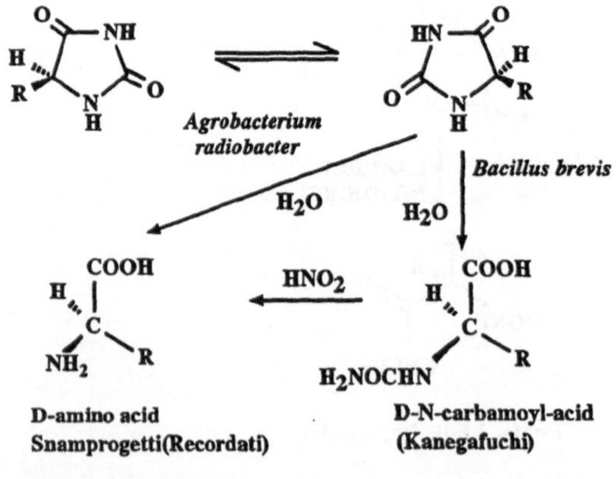

Fig. 9. Enzymatic processes for the production of optically active α-amino acids via resolution of the racemic hydantoins

firm Kanegafuchi [43]. The in situ racemization of the L-hydantoin is only effective with aromatic amino acids. Aliphatic substituted hydantoins racemize very slowly under the reaction conditions and an external recycle is necessary. Recordati uses an even more elegant approach for the production of D-*p*-hydroxy-phenylglycine on an industrial scale. The microorganism *Agrobacterium radiobacter* is able to produce both the D-hydantoinase and a second enzyme, *N*-carbamoyl-D-amino acid amidohydrolase, which catalyzes the hydrolysis of *N*-carbamoyl-D-amino acid [44].

2.5 New Developments

2.5.1 Microbial Hydrolysis of Amino Acid Carbamares (see fig. 10)

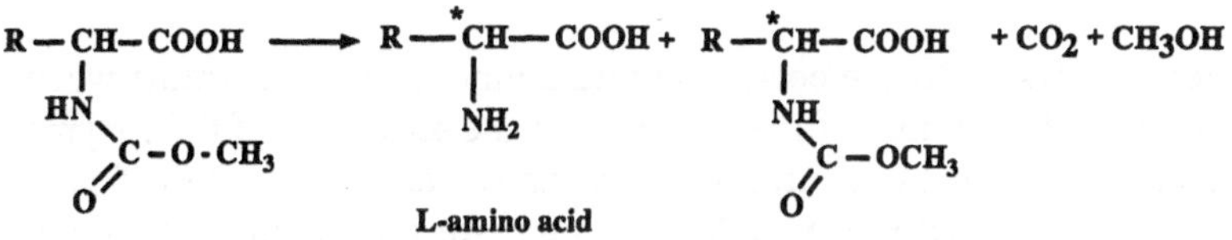

$$R-CH-COOH \longrightarrow R-\overset{*}{C}H-COOH + R-\overset{*}{C}H-COOH \quad + CO_2 + CH_3OH$$

HN	NH$_2$	NH	
C-O·CH$_3$	**L-amino acid**	C-OCH$_3$	
O		O	

Fig. 10. Microbial hydrolysis of D,L-*N*-carbamate amino acids

After a broad screening, including culture collection strains from genera known as carbamate degraders and micro-organisms from soil, Kula and coworkers [45] isolated several strains showing enzymatic activity for the stereospecific cleavage of *N*-(methoxycarbonyl)-L-alanine, *N*-(methoxycarbonyl)-L-valine and *N*-α-

Table 5. Culture collection strains exhibiting amino acid carbamate hydrolysing activity (Ref. 45)

Organism	DSM no.	Activity towards *N*-(methoxycarbonyl)-D,L-alanine		Activity towards *N*-(methoxycarbonyl)-D,L-valine	
		mU ml^{-1}	mU mg^{-1}	mU ml^{-1}	mU mg^{-1}
Agrobacterium rhizogenes	30 200	1.50	0.80	0.00	0.00
Arthrobacter globiformis	20 124	20.50	2.50	9.80	1.00
Corynebacterium sepedonicum	171	2.50	0.50	0.00	0.00
Enterobacter cloacae	30 054	0.90	0.50	0.00	0.00
Pseudomonas fluorescens	1 694	1.94	1.80	0.86	0.80
Pseudomonas fluorescens	50 124	4.99	2.35	2.99	1.65
Pseudomonas fluorescens	10 154	1.70	1.50	0.90	0.80

Table 6. Hydrolysis of N-(methoxycarbonyl)-D,L valine by different strains (Ref. 45)

Strain		Protein concentration	Substrate concentration	Incubation time	Product concentration		Yield %
		mg ml^{-1}	mM	(h)	mM	mg ml^{-1}	
A3	*Xanthomonas maltophilia*	0.2	180	65.5	18.5	2.2	33.7
A4	*Methylobacterium extorques*	1.2	180	65.0	29.5	3.5	21.1
L8	*not yet identified*	1.5	180	63.5	9.2	1.1	10.7
L10	*Pseudomonas putida*	0.5	180	67.0	46.0	5.4	52.6
L12	not yet identified	4.0	180	64.0	74.7	8.8	85.4
L18	*Alcaligenes sp.*	1.4	350	64.5	162.5	19.0	92.9

(dimethylcarbonyl)-L-lysine to the corresponding amino acids. Six strains where investigated to assess the stability of the enzyme and the selectivity of the enzyme-catalyzed conversion. From these strains *Pseudomonas fluorescens*, DSM 50 124, showed the highest specific activity with 2.3 mU mg^{-1} for the alanine and 1.6 mU mg^{-1} for the valine derivative (see Tables 5, 6).

No liberation of free amino acid could be observed upon incubating the N-(methoxycarbonyl)-D-amino acids.

The amino acid carbamates are chemically very stable. The mechanism of the enzymatic reaction observed is uncertain. One possibility is the hydrolysis of the

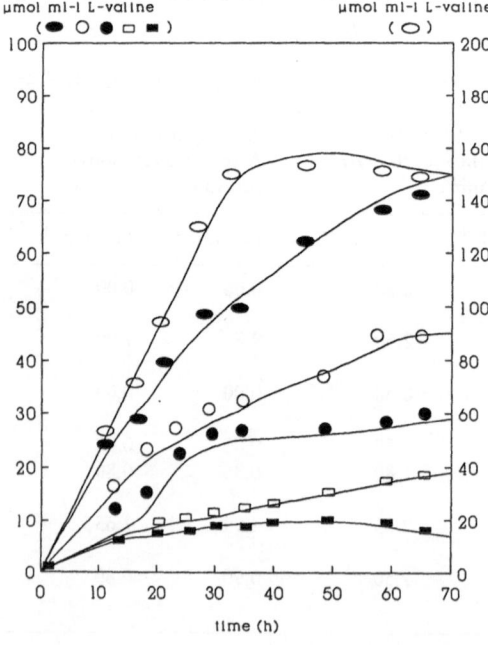

Fig. 11. Time course of the hydrolysis of N-(methoxycarbonyl)-D,L-valine with different crude enzymes

(◯) L 18, *Alcaligenes* sp.; (●) L 12, not classified; (○) L 10, *Pseudomonas putida*; (●) A 4, *Methylobacterium extorques*; (□) A 3, *Xanthomonas matophilia*; (■) L 8, not classified) (Ref. 45)

ester linkage of the amino acid carbamate leading to methanol and the corresponding substituted carbonic acid. Carbonic acid derivatives are unstable and the intermediate would presumably spontaneously decompose into the amino acid and carbon dioxide. The conversion of the reaction over 60 h was not always linear, which may reflect complex kinetics (see Fig. 11).

2.5.2 Reversal of Stereoselectivity

D-Amino-Acylase

D-Amino acids such as D-valine and D-phenylglycine are intermediates for the preparation of various pesticides, bioactive peptides and antibiotics. Known methods include the optical resolution of chemically synthesized N-acyl-D,L-amino acids with L-amino acylase (vide infra). The non-hydrolyzed N-acyl-D-amino acids are subsequently chemically hydrolyzed to obtain D-amino acids. This method was developed mainly for the purpose of producing L-amino acids; the optical purity and recovery of the D-amino acid byproducts were in general not satisfactory. For the production of D-amino acids, according to the same principle, a D-specific amino-acylase with strict stereospecificity should be a promising tool (see fig. 12).

Microbial D-amino-acylases were first reported by Kameda et al. [46, 47] in soil bacteria, by Fukugawa et al. [48–50] in *Pseudomonas* species and by Sugie et al. [51, 52] in *Streptomyces olivaceus*. These D-amino acylases have the disadvantage that their stereospecificities were not strict enough for some amino acids. Through a screening program for D-amino acylase producing micro-organisms from soil Ying-Chieh Tsai et al. [53] found a strain identified as *Alcaligenes denitrificans*, DA 181 which possesses high stereospecificity. The strain produces about 29 300 units (μmoles of product formed per hour) of D-aminoacylase and 2300 units of L-aminoacylase per gram of cells (wet weight). The specific activity of the purified enzyme was 108 600 units per mg of protein when N-acetyl-D-methionine was used as the substrate. The apparent molecular weight was 58 000. According to Table 7 N-acetyl-D-methionine is the favoured substrate, followed by N-acetyl-D-phenylalanine. The enzyme has a high stereospecificity and the hydrolysis of N-acetyl-L-amino acids was almost negligible. Until now this D-aminoacylase is the most superior in its stereospecificity and specific activity.

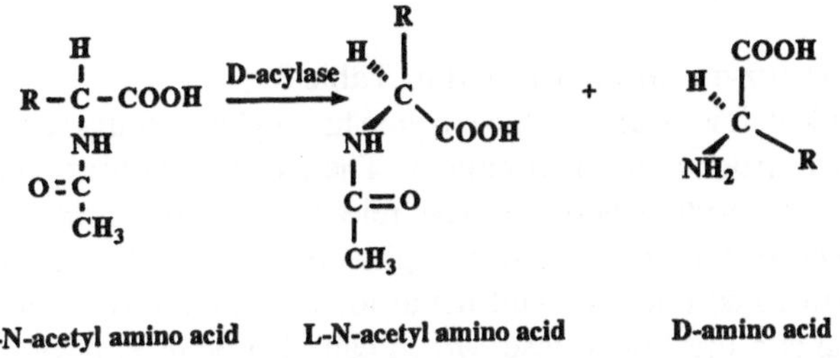

D,L-N-acetyl amino acid   L-N-acetyl amino acid   D-amino acid

Fig. 12. Reaction scheme for the kinetic resolution of D,L-N-acyl amino acids with a D-acylase

Table 7. Substrate specificity of D-amino acylase[a] (Ref. 53)

Substrate	Relative activity (%) of the[b]	
	D-form	L-form
N-Acetyl-methionine	100	0.1
N-Acetyl-phenylalanine	81	0.3
N-Chloroacetyl-valine	66	0.0
N-Acetyl-leucine	60	0.4
N-Acetyl-alanine	25	0.8
N-Acetyl-tryptophan	33	0.6
N-Acetyl-asparagine	17	0.0
N-Acetyl-alloisoleucine	12	ND[c]
N-Acetyl-valine	6	0.0
N-Acetyl-phenylglycine	5	ND
N-Acetyl-lysine	ND	0.0
N-Acetyl-aspartic acid	ND	0.0
N-Acetyl-tyrosine	ND	0.0
N-Acetyl-arginine	ND	0.0
N-Acetyl-glutamic acid	ND	0.0
N-Acetyl-histidine	ND	0.0

[a] Various substrates (20 mM) were incubated with an appropriate amount of the enzyme for 20 min at 37 °C in buffer A;
[b] The activity obtained with N-acetyl-D-methionine was assigned a value of 100;
[c] ND. Not determined

D-Stereospecific Aminopeptidase

At the Sagami Research Institute Asano et al. [54] were successful, by using an enrichment culture technique, in selecting from a soil sample a micro-organism (*Ochrobactrum anthropi*) with D-aminopeptidase activity. The enzyme, which hydrolyses D-alanine amide, was purified about 2800 fold. The molecular weight of the native enzyme was estimate to be approximately 122000 with two identical subunits having molecular weights of about 59.000.

Figure 13 shows a typical course of D-stereospecific hydrolysis of D,L-alanine amide to yield D-alanine. The rate of hydrolysis of L-alanine was less than 0.01% that of D-alanine amide.

Other substrates for the enzyme are illustrated in Table 8.

From the results in Table 8, it is clear that the enzyme has higher affinity towards peptide substrates than towards amino acid amides. The enzyme showed neither endopeptidase nor carboxypeptidase activity. The release of amino acids from D-alanylglycine in time shows a mode of action typical of aminopeptidases (EC 3.4.1.1). The effect of various compounds and metal ions on the enzyme activity indicates that the enzyme is a thiol-peptidase, which can also be used in organic solvent to stereoselectively synthesize D-amino acid N-alkylamides (see Fig. 14).

Table 8. Substrate specificity of D-aminopeptidase

Substrate	Rel. velocity in %	Km (mM)
D-alanine amide	100	0.65
glycine amide	44	22.3
D-serine amide	29	27.0
D-α-aminobutyric acid amide	30	18.3
D-alanine-3-amide-aminopentane	32	2.27
D-alanine-p-nitroanilide	96	0.51
D-alanine methylester	75	
Alanine (dimer)	21	10.2
trimer	92	0.57
tetramer	98	0.32
D-alanylglycine	95	0.98
D-alanylglycylglycine	45	0.37
D-alanyl-L-alanyl-L-alanine	100	0.65

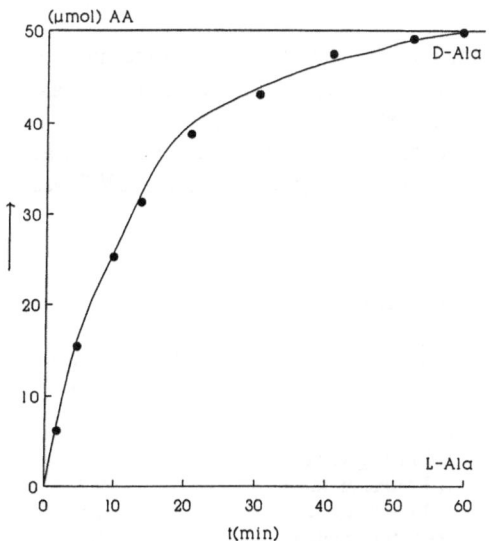

Fig. 13. Kinetic resolution of D,L-alanine amide by *Ochrobactrum anthropi* D-alanine aminopeptidase (Ref. 54)

D,L Ala-OCH₃ $\xrightarrow[\substack{\text{D-aminopeptidase} \\ \text{org. solvent}}]{\text{R - NH}_2}$ D-AlaNHR + L-Ala-OCH₃

                                              > 99 % e.e.

R = 3-pentyl, neopentyl, benzyl, n-butyl.

org. solvent water - saturated butylacetate

                          benzene

                          1,1,1, trichloroethane

Fig. 14. Stereoselective aminolysis to prepare D-amino acid *N*-alkylamides

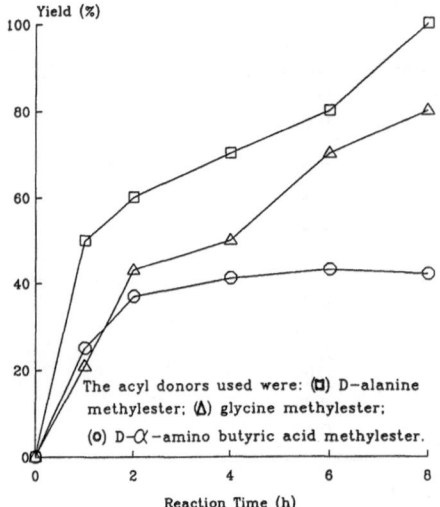

The acyl donors used were: (☐) D–alanine methylester; (▲) glycine methylester; (o) D-α–amino butyric acid methylester.

Fig. 15. Yield of the reaction with different acyl donors

The restriction of the latter system is indicated by the fact that the methylesters of D-valine, D-leucine, D-serine, D-threonine and D-methionine did not serve as a substrate in the aminolysis reaction (see Fig. 15).

A subsequent report [56] showed that several other amino acid amides also are substrates for the D-aminopeptidase from *Ochrobactrum anthropi* SV 3 (see Table 9).

Table 9. Substrate specificity of D-amino acid amidase (Ref. 56)

Substrate[a]	Relative activity (%)	Substrate	Relative activity (%)
D-Phenylalanine amide	100	D-Phenylglycine amide	15
D-Tyrosine amide	97	D-Proline amide	9.7
D-Tryptophan amide	96	D-Lysine amide	2.5
D-Leucine amide	46	D-Histidine amide	1.5
D-Norleucine amide	40	D-Asparagine	1.4
D-Alanine amide	33	D-Glutamine amide	1.1
D-Methionine amide	28	D-Phenylalanine	192
D-Norvaline amide	15	methylester	

[a] D-Threonine amide, glycine amide, D-glutamine, D-asparagine amide, D-α-amino butyric acid amide, D-isoglutamine, D-arginine amide, D-valine amide, and D-isoleucine amide were hydrolyzed by 0.1 to 0.6% the rate for D-phenylalanine amide. The following compounds were not the substrate for the enzyme: 2-Phenylacetamide, acetamide, propionamide, Boc-D-alanine amide, D-iso-asparagine, D-alanine-p-nitroanilide, D-alanylglycine, D-alanyl-D-alanine, D-alanyl-D-alanyl-D-alanine, D-alanyl-L-alanine, D-alanyl-L-alanyl-L-alanine, L-alanyl-D-alanine, L-phenylalanine amide, L-tyrosine amide, L-tryptophan amide, L-leucine amide, L-alanine amide, L-methionine amide, L-phenylglycine amide, L-proline mide, L-lysine amide, L-histidine amide, L-alanine-p-nitroanilide, L-alanyl-L-alanine, and L-phenylalanine methylester

Please note that this system was not optimal for D-alanine, but in this table D-phenylalanine is the high-activity substrate. Remarkably, D-valine amide is hydrolyzed very slowly. The M_r of the enzyme was estimated to be about 38000. The enzyme did not act on L-amino acid amides at rates faster than 0.1% of that for the corresponding D-enantiomers. The apparent K_m values for D-Phe-NH$_2$, D-Tyr-NH$_2$, D-Leu-NH$_2$ and D-Ala-NH$_2$ were calculated to be 0.089, 0.18, 0.057 and 0.54 mM respectively. The enzyme showed maximal activity at pH = 7.5.

L-*Hydantoinases*

While the enzymatic cleavage of D,L-5-monosubstituted hydantoins to D-*N*-carbamoylamino acids and D-amino acids is widespread in nature — the D-hydantoinase is identical to the dihydropyrimidinase, which plays an important role in the metabolism of pyrimidines — only a few micro-organisms are able to produce L-amino acids from D,L-5-monosubstituted hydantoins. Table 10 gives a survey of the natural occurrence of both the L-specific and the D-specific enzyme systems capable of hydrolysing D,L-5-monosubstituted hydantoins [57].

Wagner and coworkers report on the quantitative conversion of L-tryptophan from D,L-5-indolylmethylhydantoin with resting cells of a *Arthrobacter* species, DSM 3747. The optimal pH was 8.5–9.0 and the optimal temperature was 50 °C for this enzyme [58].

Yokozeki et al. [59] studied the reaction conditions for the production of L-tryptophan from D,L-indolylmethylhydantoin by a *Flavobacterium* species, AJ-

Table 10. D- and L-hydantoinases found in nature (Ref. 57)

D-Hydantoinase: in Bacteria	*Aerobacter cloaceae*
	Agrobacterium rhizogenes
	Corynebacterium sepedonicum
	Mycobacterium smegmatic
	Nocardia corallina
	Pseudomonas striata
D-Hydantoinase: in Actinomycetes	*Streptomyces almquistii*
	Streptomyces griseus
	Actinoplanes philippiensis
D-Hydantoinase: in Yeasts	*Candida utilis*
	Rhodotorula gracilis
	Torulopsis utilis
	Pichia vini
D-Hydantoinase: in Fungi	*Aspergillus niger*
in plant and animal enzyme extracts	
L-Hydantoinase: in Bacteria	*Clostridium oroticum*
	Flavobacterium ammonigenes
	Flavobacterium species
	Bacillus brevis

Table 11. Substrate specificity of the hydantoinase/decarbamoylase from *Flavobacterium* sp. AJ-3940 (Ref. 59)

Substrates	Amino acids produced	(mg ml^{-1})	Rel. value (molar base)
L-5-Indolylmethylhydantoin	L-Trp	3.80	100
L-5-(5-Hydroxyindolylmethyl)-hydantoin	5-Hydroxy-L-tryptophan	0.50	12
L-5-Benzylhydantoin	L-Phe	3.32	108
L-5-(p-Hydroxybenzyl)hydantoin	L-Tyr	2.40	71
L-5-(3,4-Dihydroxybenzyl)hydantoin	L-3,4-Dihydroxyphenylalanine	3.85	105
D,L-5-(3,4-Methylenedioxybenzyl)-hydantoin	L-3,4-Methylenedioxyphenyl-alanine	4.38	112
D,L-5-(3,4-Dimethoxybenzyl)hydantoin	L-3,4-Dimethoxyphenyl-alanine	0.15	4
Hydantoin	Gly	0	0
L-5-Methylhydantoin	L-Ala	0	0
L-5-sec-Butylhyndantoin	L-Ile	0	0
L-5-Carboxymethylhydantoin	L-Asp	0	0
L-5-Carboxyethylhydantoin	L-Glu	0	0
L-5-Methylthioethylhydantoin	L-Met	0.07	2

3940. The optimal pH of this reaction was around 8.5 and the optimal temperature was between 45 and 55 °C. The amount of L-tryptophan produced was remarkably increased by the addition of inosine to the reaction mixture — which forms a water insoluble adduct with L-tryptophan — due to the release of end-product inhibition by L-tryptophan. This inducible enzyme was intracellularly produced. Under the best conditions, 43 mg ml^{-1} of L-tryptophan was produced from 50 mg ml^{-1} of D,L-5-indolylmethylhydantoin with a molar yield of 97%. In addition, other L-aromatic amino acids such as L-phenylalanine, L-tyrosine, L-3,4-dihydroxyphenylalanine and related L-amino acids were also produced from the corresponding 5-substituted hydantoins by this bacterium (see Table 11). The same authors also reported a study on the mechanism of this reaction [60]. They concluded that the enzymatic hydrolysis of aromatic amino acid hydantoins (AAH) by *Flavobacterium* species AJ-3912 consisted of the following two succesive reactions; a hydrolytic ring opening reaction of D,L-AAH to the racemic

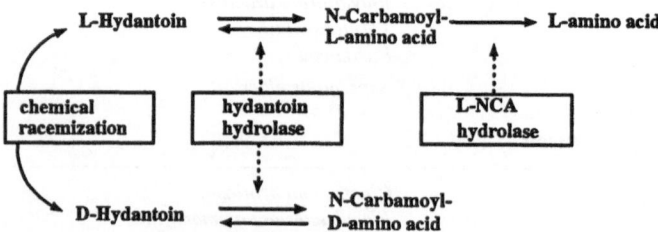

Fig. 16. Proposed scheme for the asymmetric production of L-aromatic amino acids from the corresponding hydantoins by *Flavobacterium* sp. AJ-3912

Table 12. K_m and V_{max} values of hydantoin hydrolases for 5-benzylhydantoin and N-carbamoylphenylalanine

Reaction	Km (mM)	V_{max} (mM h^{-1})
(Regular)		
L-5-Benzylhydantoin → N-Carbamoyl-L-phenylalanine	2.67	3.62
(Reverse)		
N-Carbamoyl-L-phenylalanine → L-5-Benzylhydantoin	3.13	0.66
N-Carbamoyl-D-phenylalanine → D-5-Benzylhydantoin	1.25	0.28

D,L-N-carbamoyl aromatic amino acids (NCA), catalyzed by an enzyme (hydantoin hydrolase), followed by a stereoselective hydrolytic cleavage of the L-form NCA to the L-aromatic amino acid involving another enzyme (N-carbamoyl-L-aromatic amino acid hydrolase, abbreviated as L-NCA hydrolase). The ring opening reaction involving hydantoin hydrolase was not stereospecific, but the NCA cleaving reaction involving L-NCA hydrolase was completely L-specific. The pathway for the conversion of the by-product D-NCA to L-aromatic amino acids was as follows; conversion of the D-AAH to L-AAH through spontaneous racemization, followed by the subsequent hydrolysis of the L-AAH to L-aromatic amino acids by hydantoin hydrolase and L-NCA hydrolase (see Fig. 16).

Several physical properties of the enzyme were determined, illustrative are the K_m an V_{max} values given in Table 12.

The substrate specificity of the hydantoin hydrolase and L-NCA hydrolase of *Flavobacterium* species AJ-3912 are given in Table 13 and Table 14 respectively. In the literature, most papers on screening of micro-organisms for the production of L-hydantoinases are found from around 1968 up to 1976 and from 1984 onwards. This dip in research activities with respect to L-hydantoinases is also observed in the patent literature [61]. Around 1976 it was generally believed that those enzymes were for too unstable to be used on a large production scale.

A revival in the study of L-hydantoinases is observed after 1984. Tanabe [62] reported the use of *Flavobacterium* SP1–3 with benzyl-hydantoin resulting in the formation of L-Phenylalanine. In fact, cells of *Flavobacterium* SP1–3 contain hydantoinase and decarbamoylase activities, since the L-amino acid can be directly isolated. Out of 1.5 g of D,L-5-benzylhydantoin, 1.1 g of L-phenylalanine (yield = 84%) is obtained. The micro-organism was also able to produce L-2-amino-4-phenylbutyric acid (L-homophe) from D,L-5-phenethylhydantoin [63].

Schering AG [64] reported the formation of L-amino acids from D,L-hydantoin derivatives using *Nocardia* sp. DSM 3306. Out of 10 mg 5-(2-methylpropyl)-hydantoin at 30 °C, pH = 8.5 for 24 h, 7.9 mg of L-Leucine was obtained. Ajinomoto claims that *A. Bacillus*, AJ-12299 [65] is capable of producing L-amino acids out of hydantoins. 1 g dl^{-1} D,L-5-isopropylhydantoin at 30 °C for 48 h at pH 7.5 gives rise to the formation of L-valine (0.21 g dl^{-1}).

Recently, Wagner's group has done some outstanding research in this field [66].

Table 13. Substrate specificity of hydantoin hydrolase of *Flavobacterium* sp. AJ-3912 (Ref. 62)

Substrates	Products	Relative amount of products		Ratio of products $(L/L + D)$
		(L)	(D)	
L-5-Benzylhydantoin	N-Carbamoylphenylalanine	100	0	—
D-5-Benzylhydantoin	N-Carbamoylphenylalanine	0	17	—
D,L-5-Benzylhydantoin	N-Carbamoylphenylalanine	40	11	0.78
D,L-5-(p-Hydroxybenzyl)hydantoin	N-Carbamoyltyrosine	31	4	0.89
D,L-5-Indolylmethylhydantoin	N-Carbamoyltryptophan	48	1	0.98
D,L-5-Benzyloxymethylhydantoin	N-Carbamoyl-O-benzylserine	2	19	0.10
D,L-5-(3,4-Methylenedioxybenzyl)-hydantoin	N-carbamoyl-3,4-methylene-dioxphenylalanine	11	3	0.78
D,L-5-(3,4-Dimethoxybenzyl)-hydantoin	N-Carbamoyl-3,4-dimethoxy-phenylalanine	0.7	Tr	—
D,L-5-Methylthioethylhydantoin	N-Carbamoylmethionine	Tr	Tr	—
D,L-5-Isobutylhydantoin	N-Carbamoylleucine	Tr	Tr	—
D,L-5-Isopropylhydantoin	N-Carbamoylvaline	Tr	Tr	—
D,L-5-Carbamoylethylhydantoin	N-Carbamoylglutamine	Tr	Tr	—
D,L-5-Carbamoylmethylhydantoin	N-Carbamoylasparagine	Tr	Tr	—
D,L-5-Methoxymethylhydantoin	N-Carbamoyl-O-methylserine	Tr	Tr	—
Hydantoin[a]	N-Carbamoylglycine	Tr	Tr	—
D,L-5-Methylhydantoin	N-Carbamoylalanine	0	0	—
D,L-5-Carboxymethylhydantoin	N-Carbamoylaspartic acid	0	0	—
D,L-5-Carboxymethylhydantoin	N-Carbamoylglutamic acid	0	0	—
D,L-5-(4-Imidazolmethyl)hydantoin	N-Carbamoylhistidine	0	0	—
D,L-5-(4-Aminobutyl)hydantoin	N-Carbamoyllysine	0	0	—

Tr, trace (0.2>)
[a] N-Carbamoylglycine was not formed from hydantoin

2.5.3 Racemases

In principle the biotransformation procedures mentioned above (which are kinetic resolutions) will always result in a 50% conversion of the substrate into the product leaving 50% of the substrate unchanged. A major advantage of this approach is that one obtains both enantiomers in one and the same biotransformation step. Normally, however, the markets for both the enantiomers are not in balance and recycling procedures (vide supra) have to be developed to obtain a quantitative conversion into one of the enantiomers. Therefore it would be profitable if a (constructed) micro-organism was available combining two enzyme activities, a stereoselective hydrolase in combination with a racemase capable of racemizing the unwanted (other) isomer. In fact such a situation — although it involves a chemical racemization — is already in operation in the application of D-hydantoinases for the preparation of D-p-hydroxyphenylglycine. Under the reaction conditions the aromatic L-hydantoin racemizes spontaneously, resulting in the quantitative formation of D-p-hydroxyphenylglycine. In all other cases it would be advantageous if a racemase operates in combination with a

Table 14. Substrate specificity of L-NCA Hydrolase of *Flavobacterium* Sp. AJ-3912

Substrates	Products	Relative amount
N-Carbamoyl-D,L-3,4-methylenedioxyphenyl-alanine	L-3,4-Methylenedioxyphenyl-alanine	100
N-Carbamoyl-L-phenylalanine	L-Phenylalanine	82
N-Carbamoyl-L-tyrosine	L-Tyrosine	59
N-Carbamoyl-L-tryptophan	L-Tryptophan	55
N-Carbamoyl-D,L-3,4-dimethoxyphenyl-alanine	L-3,4-Dimethoxyphenylalanine	24
N-Carbamoyl-L-methionine	L-Methionine	24
N-Carbamoyl-D,L-O-benzylserine	L-O-Benzylserine	15
N-Carbamoyl-D,L-O-methylserine	L-O-Methylserine	13
N-Carbamoyl-L-serine	L-Serine	5
N-Carbamoylglycine	Glycine	5
N-Carbamoyl-L-leucine	L-Leucine	3
N-Carbamoyl-β-alanine	β-Alanine	3
N-Carbamoyl-L-isoleucine	L-Isoleucine	2
N-Carbamoyl-L-valine	L-Valine	2
N-Carbamoyl-L-glutamine	L-Glutamine	1
N-Carbamoyl-L-asparagine	L-Asparagine	1
N-Carbamoyl-L-alanine	L-Alanine	0.5 >
N-Carbamoyl-L-glutamic acid	L-Glutamic acid	0
N-Carbamoyl-L-aspartic acid	L-Aspartic acid	0
N-Carbamoyl-L-lysine	L-Lysine	0
N-Carbamoyl-L-histidine	L-Histidine	0
N-Carbamoyl-L-proline	L-Proline	0

stereospecific hydrolase. For example, a N-acylamino acid racemase in combination with the L- or D-amino-acylase, an aliphatic hydantoin racemase in combination with a L- or D-specific hydantoinase, an amido-amino acid racemase in addition to a stereospecific L- or D-decarbamoylase and an amino acid amide racemase in combination with a peptidase.

Amino acid racemases have long been known. Since the discovery of alanine racemase several amino acid racemases catalyzing interconversions between D- and L-amino acids have been found in bacteria and some of them have been characterized [67, 68]. Evidently, no amino acid racemases should be present in the applications discussed above.

Recent results on the use of racemases other than amino acid racemases are given in this review.

N-Acylamino Acid Racemases

Researchers at Takeda [69] obtained an acylamino acid racemase through screening. Acylamino acid racemase-producing micro-organisms were obtained by selecting strains which do not exhibit amino acid racemase activities and which convert D-N-α-acylamino acid into the corresponding L-α-amino acid. The N-

Table 15. Acylamino acid racemase producing micro-organisms found through screening (Ref. 69)

Acylamino Acid Racemase producing micro-organism	Production of L-methionine from N-acetyl-D-methionine (25 mM)	Production of D-methionine from L-methionine (25 mM)
Streptomyces sp. Y-53 (FERM P-9518)	12.7 mM	0 mM
Actinomadura roseoviolacea (IFO 14098)	1.2	0
Actinomyces aureomonopodiales (IFO 13020)	1.6	0
Jensenia canicruria (IFO 13914)	2.1	0
Amycolatopsis orientalis (IFO 12806)	1.4	0
Sebekia benihana (IFO 14309)	3.9	0
Streptomyces coelescens (IFO 13378)	4.2	0
Streptomyces cellulflavus (IFO 13780)	3.2	0
Streptomyces alboflavus (IFO 13196)	2.9	0
Streptomyces aureocirculatus (IFO 13018)	2.3	0
Streptomyces diastatochromogenes (IFO 13389)	1.8	0
Streptomyces spectabilis (IFO 13424)	2.8	0
Streptomyces tuirus (IFO 13418)	3.8	0
Streptomyces griseoaurantiacus (IFO 13381)	2.7	0

acylamino acid racemase-producing micro-organisms which have been found by this method are listed in Table 15.

The enzyme is a true racemase, i.e. the L-N-α-acylamino acid is converted into the corresponding D-N-α-acylamino acid, and vice versa. The substrate specificity seems to be rather relaxed as is illustrated in Table 16. Moreover, as shown in this

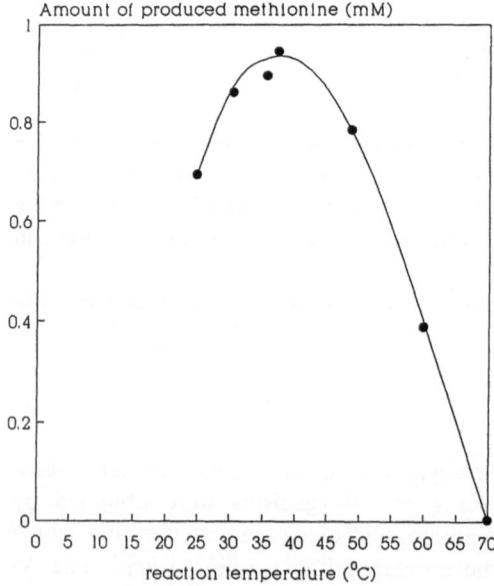

Fig. 17. The relationship between reaction temperature and enzyme activity in acylamino acid racemase (Ref. 69)

Table 16. Substrate specificity of the *N*-acylamino acid racemase from *Streptomyces* sp. Y-53 (Ref. 69)

Substrate	Relative activity
N-acetyl-D-methionine	100
N-formyl-D-methionine	40
N-acetyl-D-alanine	33
N-benzoyl-D-alanine	14
N-acetyl-D-leucine	37
N-acetyl-D-phenylalanine	64
N-chloroacetyl-D-phenylalanine	90
N-acetyl-D-tryptophan	10
N-acetyl-D-valine	35
N-chloroacetyl-D-valine	80
N-acetyl-D-alloisoleucine	33
D-methionine	0
D-alanine	0
D-leucine	0
D-phenylalanine	0
D-tryptophan	0
D-valine	0
N-acetyl-L-methionine	100
N-formyl-L-methionine	63
N-acetyl-L-alanine	21
N-benzoyl-L-alanine	ND
N-acetyl-L-leucine	74
N-acetyl-L-phenylalanine	84
N-chloroacetyl-L-phenylalanine	112
N-acetyl-L-tryptophan	80
N-acetyl-L-valine	19
N-chloro-acetyl-L-valine	105
N-acetyl-L-alloisoleucine	ND
L-methionine	0
L-alanine	0
L-leucine	0
L-phenylalanine	0
L-tryptophan	0
L-valine	0

ND: Not determined

table, the enzyme acts only on optically active *N*-acylamino acids but does not act on the corresponding optically active α-amino acids.

The temperature range most suitable for the activity of the combined enzyme system (racemase and L-amino acylase) has also been studied. The amount of L-methionine produced from *N*-acyl-D-methionine was determined within the reaction temperature range of 25–70 °C. The amount of enzyme used was increased and the reaction time was kept constant at 5 min. As is evident from Figure 17 the suitable temperature range for the activity is from 30 to 50 °C.

The enzyme-stability as a function of temperature is shown in Fig. 18. As is evident from the curve, the enzyme is stable below 40 °C but inactivates above

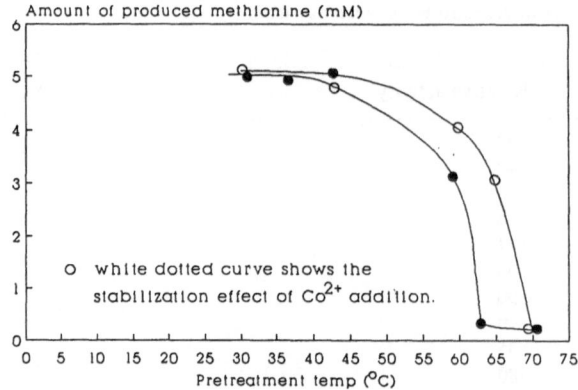

Amount of produced methionine (mM)

o white dotted curve shows the
   stabilization effect of Co^{2+} addition.

Pretreatment temp (°C)

Fig. 18. The heat stability of the acylamino acid racemase (Ref. 69)

50 °C. A stabilization effect is observed upon the addition of Co^{2+} as shown in Fig. 18 (white dotted curve). In the presence of Co^{2+} the enzyme is stable below 50 °C and inactivates at 60 °C.

The influence of metal-ions was also studied as is illustrated in Table 17.

As shown in Table 17, the enzyme is noticeably activated within certain limited ranges of concentrations of several kinds of metal ions. Cobalt ions have a noti-

Table 17. Influences of metal ions on enzyme activities (Ref. 69)

Influences of Metal Ions on Enzyme Activities

Additive	Additive Concentration	
	1 mM	10 mM
None	100	100
LiCl	100	120
NaCl	100	130
KCl	110	130
$MgSO_4 \cdot 7 H_2O$	890	1350
$CaCl_2$	95	100
$SrCl_2 \cdot 6 H_2O$	94	110
$BaCl_2 \cdot 2 H_2O$	85	120
$Al_2(SO_4)_3 \cdot 16 H_2O$	79	0
$CrCl_3 \cdot 6 H_2O$	96	100
$MnSO_4 \cdot 7 H_2O$	780	1050
$FeSO_4 \cdot 7 H_2O$	100	560
$CoCl_2 \cdot 6 H_2O$	2100	140
$NiSO_4 \cdot 6 H_2O$	460	0
$CuSO_4 \cdot 5 H_2O$	0	0
$ZnSO_4 \cdot 7 H_2O$	810	80
$NaMoO_4 \cdot 2 H_2O$	96	0
$SnSO_4$	92	100
$PbCl_2$	78	100
EDTA	68	0

ceable activating effect at low concentrations (around 1 mM), while copper ions showed a noticeable inhibitory effect. Aluminium ions and molybdic ions also inhibit the enzyme activity at 10 mM.

N-Carbamoyl-α-Amino Acid Racemase

The conversion of D-hydantoins into L-amino acids has been described [70]. Through screening a micro-organism from the species *Coryneforme* containing three enzyme activities, i.e. racemase activity for *N*-carbamoyl-α-amino acids, enantioselective L-*N*-carbamoyl-α-amino acid amido-hydrolase and a non-selective hydantoinase was obtained (Table 18).

Table 18. Substrate specificity of species *Coryneforme*

Substrate 5-substituent	*N*-carbamoyl amino acid		α-amino acid	
	% D	% L	% D	% L
Ethyl-2-methylthioether-	100	0	0	100
Benzyl-	50	50	0	100
p-Hydroxybenzyl-	56	44	0	100
3,4-dihydroxybenzyl	58	42	0	100
Benzylmethylene-ether-	50	50	0	100
3-Methylenindol	6	94	0	100

In Table 19 the conversion of L-, D,L- and D-*N*-carbamoyl amino acids into the corresponding L-amino acids is indicated, clearly demonstrating the presence of a racemase.

Table 19. Conversion of D,L-, L- and D-*N*-carbamoyl amino acids (Ref. 70)

Substrate	
D,L-*N*-carbamoyltryptophane	L-Tryptophan
L-*N*-carbamoyltryptophane	L-Tryptophan
D,L-*N*-carbamoylphenylalanine	L-Phenylalanine
D-*N*-carbamoylphenylalanine	L-Phenylalanine
D-*N*-carbamoylmethionine	L-Methionine

The enzyme is active at 20–50 °C and pH 6.5–10

Amino Acid Amide Racemases

Since the aminopeptidase resolution process discussed in Sect. 2.3 can be significantly improved with the application of amino acid amide racemases, an extensive screening program was launched. Unfortunately, no generally applicable amino acid amide racemases could be found. Therefore, a different approach, comprising in vivo protein engineering techniques, was investigated. This approach is discussed in detail in Sect. 3.

3 Methodologies to Obtain Novel Enzyme Activities

A wider introduction of biocatalysts in industrial chemo-enzymatic processes is severely hampered for the following reasons:

1. A lack of operational stability, i.e. relatively low stability under process conditions. The major factors required for improvement of the productivity of microbial whole cell biocatalysts or enzymes are the resistance to: product inhibition, high substrate and salt concentrations, protease inactivation, organic solvents, low or high pH-values, higher temperatures;
2. A generally high substrate specificity, which limits the application of the biocatalyst towards a broad range of comparable substrates.
3. A shortage of biocatalysts with a high degree of enantiospecificity, required for the production of chiral products.

Moreover, the screening and development of new biocatalysts with specific (unnatural) and commercially interesting activities, necessary for further application of biocatalysts in industrial production processes, is restricted because of a limited availability and variety of goal-orientated screening methods [71]. Except for a few specific examples most approaches in screening strategies are general and non-specific. These exceptions mostly refer to the natural habitat of the microbial species used in the screening (e.g. thermophiles for more stable enzyme activities at high temperature), or to a specific compound that needs to be degraded and therefore can be supplied as the only carbon and/or nitrogen source to the media, or refer to the production of a specific compound which can be easily detected by fast chemical or (micro)biological methods.

The use of hydrolases is already well-established in the area of amino acid synthesis and has provided the pharmaceutical and agrochemical industry with an important chiral pool from which optically active products can be made (vide supra).

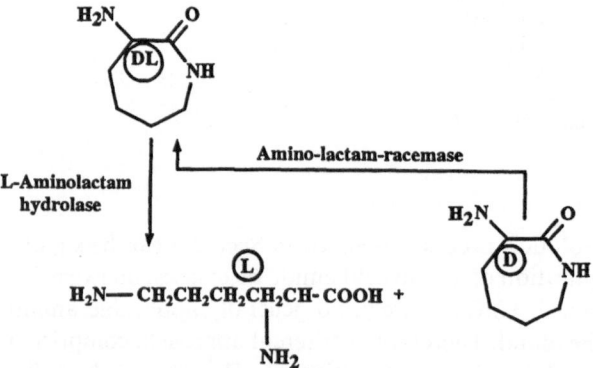

Fig. 19. Enzymatic synthetic of L-lysine from D,L-α-amino-ε-caprolactam

However, an important disadvantage of this class of enzymes in the production of chiral intermediates is the product yield: With a few exceptions this method of enzymatic kinetic resolution of racemic mixtures — without an external recycle for the unwanted isomer — does not result in a stoichiometric conversion. One of the exceptions is the production of L-lysine from D,L-α-amino-ε-caprolactam [72–74]. This process is based on the combination of two enzymatic reactions: the enzymatic enantiospecific hydrolysis of L-α-amino-ε-caprolactam to L-lysine and the simultaneous racemisation of the D-α-amino-ε-caprolactam into the L-enantiomer (Fig. 19).

In this way L-lysine is commercially produced from D,L-α-amino-ε-caprolactam with a yield of nearly 100 % by incubating the racemate with microbial cells of *Cryptococcus laurentii*, possessing L-α-amino-ε-caprolactamase activity, together with cells of *Achromobacter obae*, possessing α-amino-ε-caprolactam racemase activity.

Another option to produce optically active amino acids with a yield of 100 % is the enzymatic resolution of aromatic racemic hydantoins, e.g. for the production of L-phenylalanine, an intermediate for the artificial dipeptide sweetener aspartame, and for the synthesis of D-phenylglycine and D-*p*-hydroxyphenylglycine, intermediates for the broad spectrum antibiotics Ampicillin and Amoxycillin.

Further optimisation of the industrial processes for the production of amino acids based upon the enzymatic resolution of racemic substrates (e.g. amidase, hydantoinase, acylase and esterase technology) requires the screening of specific enzyme activities with improved characteristics concerning the operational stability. This includes the screening of further D- and L-specific hydrolase activities and the screening of different racemase activities.

The emphasis in these screening operations should be on rational direct or indirect methods including the definition of the criteria for screening of the specific enzymatic activity [71, 75, 76]. Cheetham [71] gives an example of a possible flow diagram of the procedures involved in a classical screening (Fig. 20).

The design of rational direct and indirect screening methods could involve, for example the use of auxotrophs, the reversion of non-producers and resistance to toxic analogs or relief of end-product inhibition.

An example of a such a direct method in the screening of enzymes for the production of amino acids is the screening for micro-organisms capable of producing L-amino acids from corresponding, D,L-hydantoins, described by Groß [77].

This method comprises the use of a tryptophan-auxotrophic yeast in an overlay assay to enable the detection of L-tryptophan producing micro-organisms from D,L-indolylmethylhydantoin. Together with the indirect method for enrichment and selection of hydantoinase producing micro-organisms [78], a biocatalyst can be selected with high hydantoinase activity under different process conditions. This method offers a direct screening of operationally stable biocatalyst with specific hydantoinase activity. Another direct approach is the use of the conventional techniques of batch and plate enrichment and continuous cultivation.

Genetic engineering techniques, such as protein engineering [79], also offer a promising method to obtain mutant enzymes with e.g. higher operational stability and even new classes of enzymes.

SCREENING AND OPTIMIZATION

I __Classical approach:__

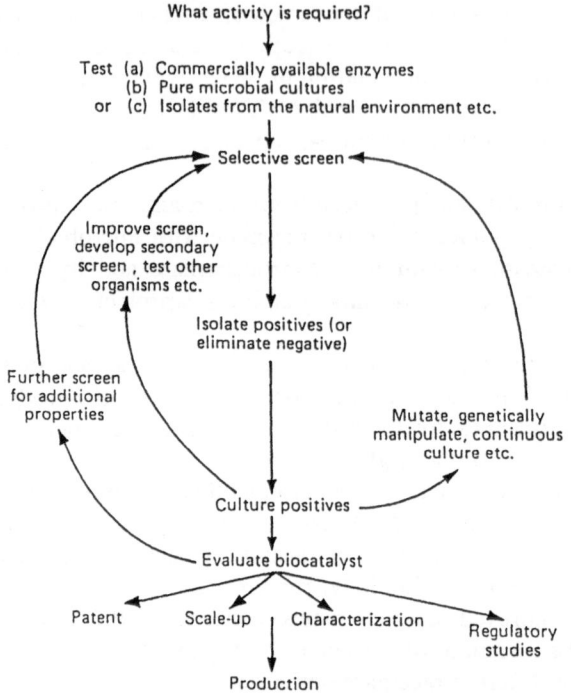

Flow diagram of important screening operations

Fig. 20. Flow diagram of important screening operations (Ref. 71)

The impact of protein engineering on the generation of optimised biocatalyst is shown by the example of Subtilisin [80]. Almost every property of this enzyme, a bacterial serine protease, has been altered by protein engineering including its catalytic activity, substrate specificity, pH optimum and stability to oxidative, thermal and alkaline inactivation. Although these techniques do not require selective culture conditions and detection methods compared with the classical screening techniques, a detailed knowledge about structure activity relations with respect to the enzyme system under investigation is absolutely necessary. Moreover, the physiological effects on the host organism by a desired mutation is not clear beforehand. Moreover, these in vitro protein engineering techniques are mainly concerned with improving upon nature using already existing molecules. Also, the limited knowledge of structure-function relations prevents the ab initio construction of new enzyme activities. So, why not return to nature and let nature do the job?

Micro-organisms are considered to be a very ancient biological group and therefore should have a very flexible and vast genetical potential to be able to survive and adapt to a large variety of habitats, expecially those created by the chemical industry in the last decades. This is the example of 'in vivo' protein engineering on a large scale. Cultivating micro-organisms in the laboratory under selective conditions, particularly by using chemostat culture, novel (and unnatural) enzyme activities and metabolic pathways can evolve, using 'the microbe as an engineer' [81, 82].

In vivo protein engineering [83–85] offers not only the possibility of exploiting the largely cryptic or silent parts of the microbial genomes [86] to generate new and commercially attractive enzyme activities, but also can be used to modify existing enzymes by changing the coding nucleotide sequences to generate enzyme activities with a higher operational stability. The exploitation of the largely 'unused' parts of the microbial genomes is based upon experimental enzyme evolution where high selective pressure, e.g. in chemostat cultures, gives rise to physiologically stable biocatalysts, expressing the desired enzyme activities.

Several investigations already studied the possibilities of in vivo protein engineering to modify the characteristics of existing enzymes and to generate new enzyme activities [87]. In one of these investigations, Hartley studied the experimental evolution of ribitol dehydrogenase and as conclusion he postulated the crucial role of gene doubling in the evolution of enzymes and the generation of a new class of enzyme activities [88] (Fig. 21).

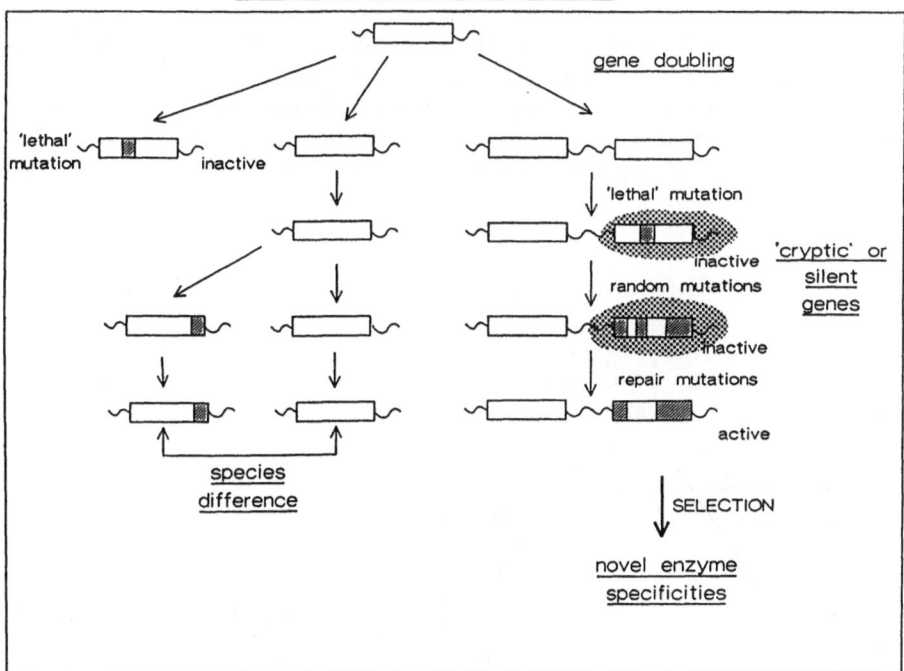

Fig. 21. Gene doubling in evolution. *Shaded areas* show mutations in the genes (according to B. S. Hartley)

The use of in vivo protein engineering for optimization of existing biocatalysts, especially with respect to the operational stability has already been proven by the selection of a thermostable variant of the kanamycine nucleotidyl transferase enzyme. This has been achieved by cloning this *Staphylococcus aureus* gene into a thermophile *Bacillus stearothermophilus* host strain. By subsequent selection at elevated temperatures under conditions where mutant strains with the desired enzyme activity have an advantage in growth, the temperature stability of this enzyme could be increased from 47 °C to 70 °C [89].

Another approach for the screening of enzyme activities with higher operational stability is the exploration of extreme environments for microbial types and activities [90].

In the course of evolution extremophile micro-organisms have adapted to life under extreme conditions: high or low temperatures, extreme pH-values, high salt concentrations, etc. Extremophile biocatalysts could have many advantages concerning the operational stability. They are known not only to be stable and active at high temperatures, but also to be more resistant to denaturing impurities and proteolytic inactivation [91]. An excellent example of the use of in vivo protein engineering for the generation of an optimised biocatalyst in the production of optically active amino acids is the optimisation of the *Pseudomonas putida* biocatalyst used in the enzymatic resolution process developed by DSM (see Sect. 2.3). The availability of a D-specific amidase activity could result in the formation of the D-amino acid without the additional hydrolysis step. If only one of the amino acid enantiomers is desired because of the market situation, racemization at the substrate level would greatly enhance the economical feasibility of the production process. Appropriate selective pressure in chemostat culture was found to result in the selection of a mutant strain with in addition to the original L-specific amidase activity, novel D-amidase and amino acid amide racemase activities [92]. The availability of a biocatalyst with amino acid amide racemase activity together

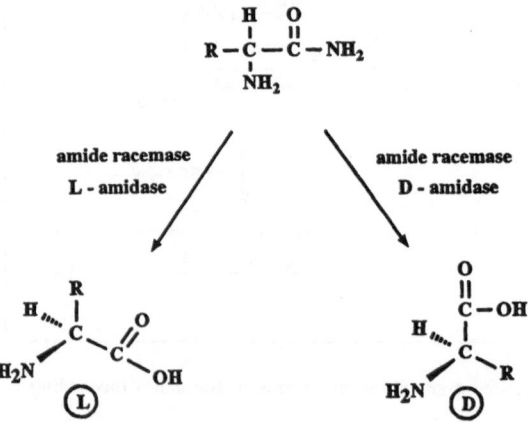

Fig. 22. The production of optically active amino acids, using a biocatalyst with amino acid amide racemase activity together with D- or L-specific amidase activity

with D- or L-specific amidase activity would result in a very efficient process for the production of optically active amino acids with a theoretical yield of 100% (Fig. 22).

The construction of a biocatalyst with a combination of these enzymatic activities out of the obtained *Pseudomonas putida* mutant strain requires the generation of respectively L- and D-amidase negative strains. The mutagenesis procedure to obtain these amidase negative mutants should be based upon the generation of stable deletion mutants, e.g. by using diepoxyoctane or transposon mutagenesis [93].

The above-mentioned examples strongly emphasize the importance of in vivo protein engineering as a goal-orientated method for the screening of commercially attractive biocatalysts and for the optimization of biocatalysts, especially with respect to the operational stability. Hopefully, the use of this screening method and the generation of new and stable biocatalyst will give a strong impetus on the further application of enzyme activities in industrial processes.

4 Approaches to the Chemo-Enzymatic Synthesis of Amino Acids Used in the Production of Artificial Dipeptide Sweeteners

4.1 L-Phenylalanine

The introduction of aspartame or L-N-aspartyl-L-phenylalanine methylester as a non-nutritional dipeptide sweetener has generated wide spread interest in the synthesis of this compound and its constituents, L-aspartic acid and L-phenylalanine. L-aspartic acid can be obtained from protein hydrolysates or from enzymatic synthesis starting from fumaric acid [94].

For phenylalanine numerous synthetic routes, based on chemical, microbial and chemo-enzymatic procedures have been explored. In Fig. 23a number of routes has been depicted, some of which have been commercialized, although recent developments in the microbial production of phenylalanine using genetically engineered micro-organisms have made most routes obsolete. Still, the chemo-enzymatic approaches nicely illustrate the potential of biocatalytic procedures and are therefore discussed in this paragraph.

Initially, L-phenylalanine was produced using the acylase-technology [95, 96], in which D,L-N-acetylphenylalanine is resolved. The substrate can be produced in a number of ways, which will not be discussed in detail. Starting materials include benzaldehyde and N-acetylglycine, styrene oxide, formamide and CO, or phenylpyruvic acid (synthesised from benzylchloride and CO, followed by a reductive amination). Alternatively, L-phenylalanine can be produced from phenylpyruvic acid using phenylalanine dehydrogenase in a stereospecific reductive amination [97]. Polyethylene glycol modified NADH can be used as the cofactor which can

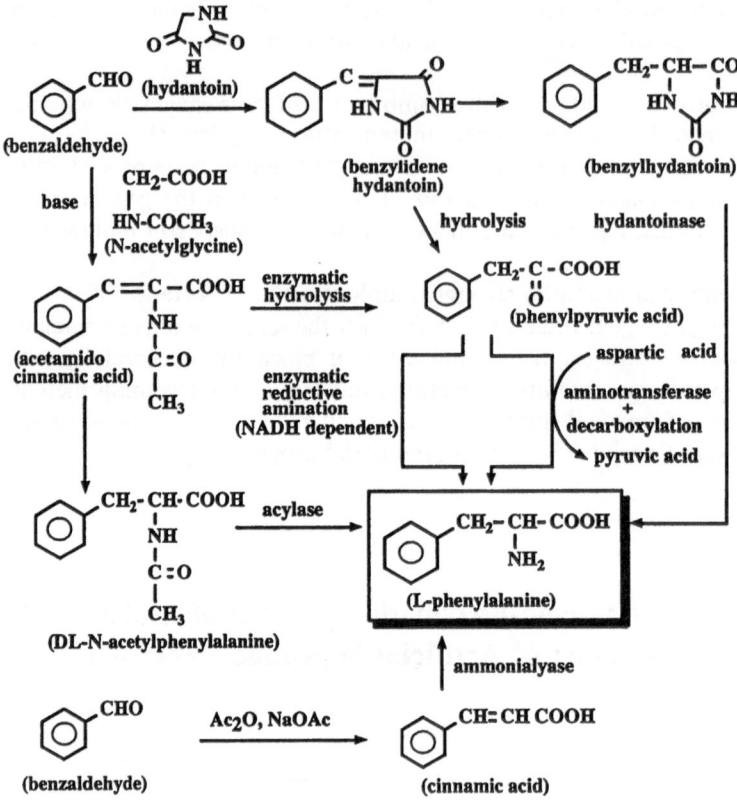

Fig. 23. Chemo-enzymatic routes for the synthesis of L-phenylalanine

be regenerated with formate dehydrogenase. However, despite this efficient co-factor regeneration system the process has never been commercialized. Phenyl-pyruvic acid can be used as starting material in the synthesis of phenylalanine in another, less complex, process using the aminotransferase technology [98]. L-As-partic acid was used as amino donor. The equilibrium was shifted towards L-phenylalanine formation by decarboxylation of the initially formed oxaloacetate under the reaction conditions. This process has been operated on a commercial scale but became obsolete due to the development of an efficient microbial process for L-phenylalanine production. Phenylpyruvic acid can be produced enzymatically from acetamido-cinnamic acid [99] or chemically via hydrolysis of benzylidenehy-dantoin obtained upon condensation of benzaldehyde and hydantoin. Benzylide-nehydantoin can be further hydrogenated to benzylhydantoin. The latter can be converted into L-phenylalanine upon treatment with a L-specific hydantoinase [59, 60]. A very elegant and at one time commercially very attractive method for L-phenylalanine production is the conversion of cinnamic acid and ammonia into

L-phenylalanine using an ammonialyase [100, 101]. This process has also been operated on a commercial scale.

De Boer and Dijkhuizen [102] have extensively reviewed microbial and enzymatic processes for L-phenylalanine production and further details can be found in their review in this series.

A techno-economic evaluation of the processes mentioned above is difficult to make, largely because virtually no free market for L-phenylalanine exists due to the dominant aspartame patent position of Monsanto-Nutrasweet. In general, however, it can be stated that on a large scale the microbial process probably is the most economic, because of the low raw material costs. Most chemo-enzymatic approaches, with the possible exception of the ammonialyase route, are based on the conversion of relatively expensive starting materials. Those processes also require a (relatively) large-scale of operation to be economic and future commercialization of those processes depends to a major extent on the market developments for aspartame.

4.2 D-Alanine

Alanine is a nonessential amino acid used on a limited scale (in L and D,L forms) in Japan as a seasoning. The largest potential use for D-alanine is as a constituent of a new sweetener, Alitame, developed by Pfizer. Alitame joins a small, but expanding, number of low-calorie and noncaloric sweeteners in the food marketplace. Alitame, is a dipeptide of L-aspartic acid and a heterocyclic substituted amide derivative of D-alanine with the following structure; L-aspartyl-D-alanine-N (2,2,4,4-tetramethylthietan-3-yl)amide. Thus it differs from the dipeptide sweetener aspartame, because of the substitution of L-phenylalanine methyl ester by a D-alanine amide derivative, a structural change said to lead to a product with improved heat stability and a 12 times more intense sweetness. A steadily growing amount of literature on the preparation of D-alanine is available. A microbial production process has been developed by the Tanabe Seiyaku Co. in Japan [103] based on *Corynebacterium fascians* (ATCC 21950), which converts glucose to D-alanine. Several classical resolution processes have been developed [104–107]. In addition, several enzymatic approaches can be envisaged, including a transamination process involving the transamination of pyruvic acid, the use of immobilized D-acylases, of D-aminopeptidases or D-hydantoinases or the use of a yeast strain capable of digesting L-alanine from the racemic mixture.

In Fig. 24a number of possible routes to D-alanine is depicted.

Although usually microbial processes are only suitable for the preparation of naturally occurring L-amino acids, there seems to be one exception, namely the production of D-alanine [108]. Alanine is the amino acid most commonly accumulated by micro-organisms and many investigations have been reported on its microbial production [109]. In most cases, the D,L form accumulates, which can be explained by the fact that alanine racemase is more widely distributed in many organisms than other amino acid racemases.

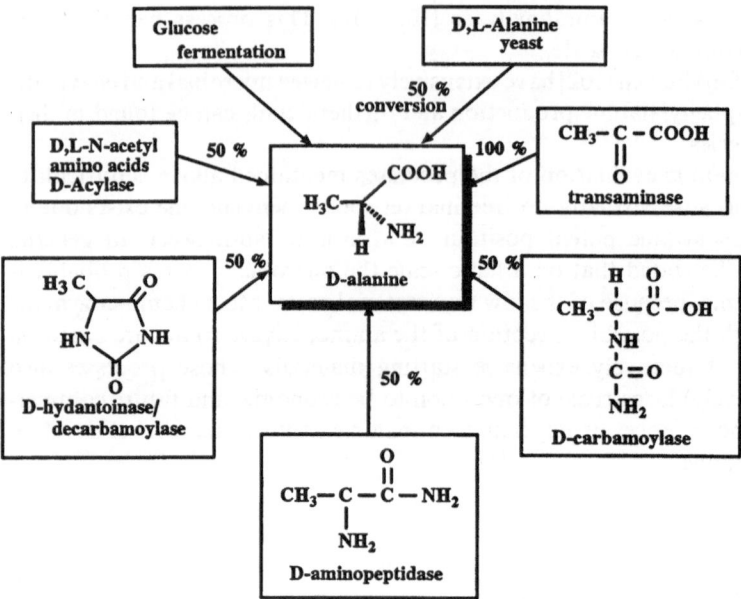

Fig. 24. Several routes for the synthesis of D-alanine

D-Alanine is a constituent of bacterial cell walls, antibiotics and tumor proteins [110].

Chibata discovered that a large amount of D-alanine is extracellularly excreted by *Corynebacterium fascians* ATCC 21 950 [108]. Corn hydrolyzate (68 weight percent glucose) is used as the glucose source in the batch microbial process of D-alanine. The medium is cultivated at 30 °C for 5 days. Yields up to 12 mg ml^{-1} are reported. The mechanism of production of D-alanine by *Corynebacterium fascians* has been discussed by Yamada et al. [111]. These workers used glycerol as the raw material for D-alanine and found that a marked increase in production resulted from the addition of pyruvate to the medium. They concluded that in the transformation, L-alanine is formed first from pyruvate. An alanine racemase was detected, which is assumed to catalyze the conversion of L-alanine to a D,L-alanine mixture. The preferential excretion of the D-form through the cell wall allows for the continuous action of the racemase, thus shifting the equilibrium in favor of the D-amino acid. In their work, the authors were able to eliminate the possibility of two other mechanisms, the transamination between pyruvate and other amino acids, and the decarboxylation of aspartic acid.

A chemo-enzymatic method to obtain a theoretical yield of 100% would be a transamination [112] of pyruvic acid (see Fig. 24). Most other chemo-enzymatic approaches are based on kinetic resolution and give 50% of the desired enantiomer. The applications of D-hydantoinases/carbamoylases, D-aminopeptidases and

D-acylases have been discussed before and will not be elaborated upon (see Fig. 24). Other approaches include the use of D-esterases [113, 114], D-decarbamoylases [115] and the use of yeast strains for the conversion of D,L-alanine into D-alanine via digestion of the L-enantiomer [116].

D,L-Esters can be hydrolyzed by conventional esterases. D-Alanine alkyl esters are obtained with a reasonable enantiomeric excess (98%), while the L-amino acid formed in this reaction is obtained with low optical purity (85%). The reverse reaction is also known. Kise et al. [114] reported the esterification of N-acetylated amino acids with α-chymotrypsin suspended in alcohols. The rate and the yield of the ester depended primarily on the amount of aqueous buffer solution in which α-chymotrypsin was dissolved. The reaction rate increased but the equilibrium yield of the esters decreased with increasing concentration of water. High substrate selectivity and stereospecificity were observed for L-aromatic amino acids. D-alanine can also be obtained by the conversion of N-carbamoyl-alanine using a D-decarbamoylase from *Pseudomonas hydantoinophilum* AJ 1120 [115].

Yeast cells, capable of digesting L-alanine but not D-alanine in acidic medium (pH = 4–6.5) can also be used for the production of D-alanine [116]. Toray Industries report that several yeast strains can be used, e.g. *Candida, Saccharomycopsis, Pichia, Torulopsis, Hansenula, Cryptococcus* or *Trichosporon*. It is stated that *Candida humicola* ATCC 36992, when incubated with $100 \, g \, l^{-1}$ of D,L-alanine affords 48 g of D-alanine after 70 h.

Potentially, the microbial process for for D-alanine production would be a technically feasible option. But for economical reasons the titre should be higher and the reaction time should be shorter. The D-amidase methodology would be effective if the L-alanine amide could be racemized easily during the reaction using an amino acid amide racemase (vide supra). The use of D-transaminase is more complicated and does not appear to be a feasible option. All other chemo-enzymatic approaches have a theoretical yield of 50% and require a separate racemization step.

5 Chemo-Enzymatic Synthesis of α-Alkyl-α-Amino Acids

A well known application of the of use α-alkyl-α-amino acids in the pharmaceutical industry is L-α-methyl-3,4-dihydroxyphenylalanine, which is used as a drug to treat patients suffering from high blood pressure [117]. More recently, medicinal chemists have become interested in bio-active peptides containing α-alkyl-α-amino acids, since they tend to freeze specific conformations and slow down dramatically enzymatic processes [118]. Nowadays, many α-alkyl-α-amino acids can be found in nature. For example L-isovaline is found in peptaibol antibiotics [119, 120]. Their influence on the conformational behaviour of peptides is presently under active investigation [121, 122]. Several routes to enantiomerically pure α-alkyl-α-amino acids have been elaborated during the last few years [123–127].

Amidases

At DSM a new biocatalyst from *Mycobacterium neoaurum* was obtained through screening, capable of stereoselectively hydrolyzing a range of α-alkyl-α-amino acid amides [20–22, 128–130]. The basis of this process is essentially the same as that for α-H-amino acids (see 2.3). In this case, however, a ketone is used as the starting material, which undergoes a Strecker reaction, followed by hydrolysis of

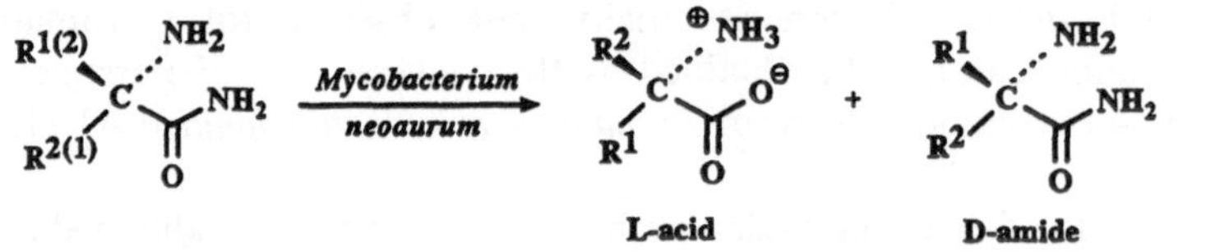

R^1	R^2	rotations, L-acids
$(CH_3)_2CH$	CH_3	$-4.0°$ (c 1.3, H_2O)
$(CH_3)_2CHCH_2$	CH_3	$+34.2°$ (c 3, H_2O)
C_6H_5	CH_3	$-86.0°$ (c 1, 1 n HCl)
$C_6H_5CH_2$	CH_3	$-22.0°$ (c 1, H_2O)
$4\text{-}CH_3OC_6H_4CH_2$	CH_3	$-6.9°$ (c 1, HCl)
$C_6H_5(CH_2)_2$	CH_3	$+38.1°$ (c 1, HCO_2H)
$C_6H_5CH_2$	C_2H_5	$-22.8°$ (c 2, H_2O)

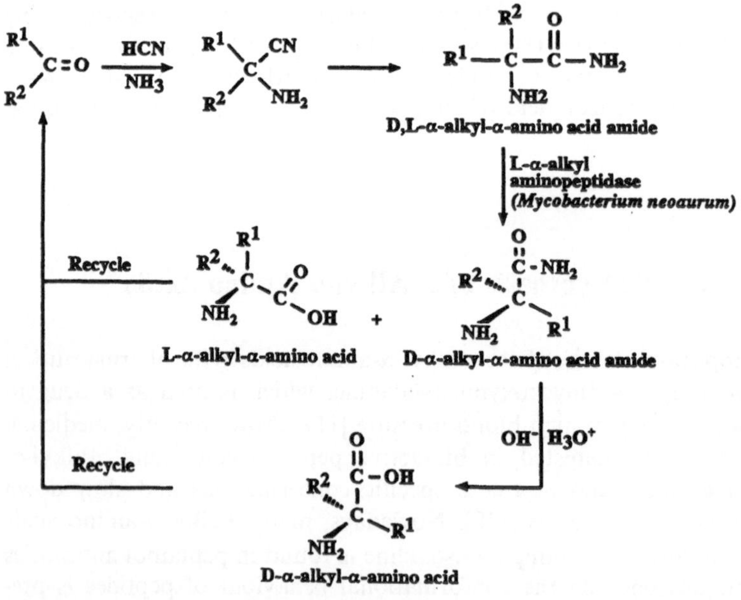

Fig. 25. DSM chemo-enzymatic resolution process for the synthesis of L- and D-α-alkyl-α-amino acids, including the optical rotation of the L-enantiomer produced (Ref. 123)

Table 20. Substrate specificity of the amidase from *Mycobacterium neoaurum*

Substrate specificity:

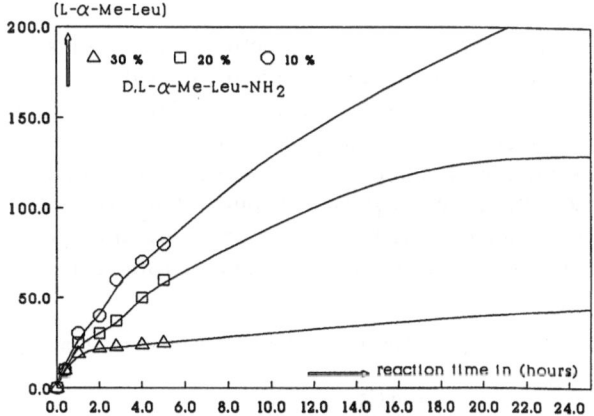

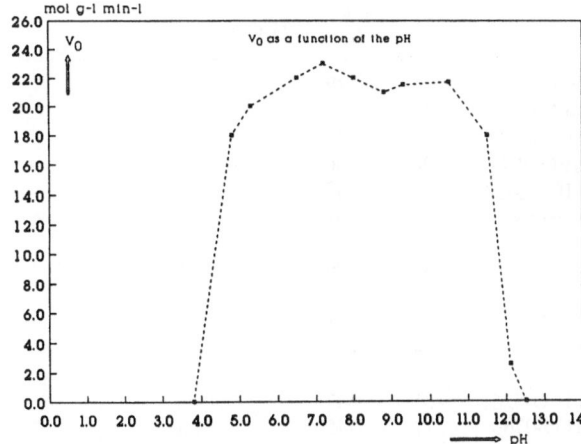

Fig. 26. Kinetics of the enzymatic resolution of D,L-α-methyl-leucine amide. The pH optima of the reaction

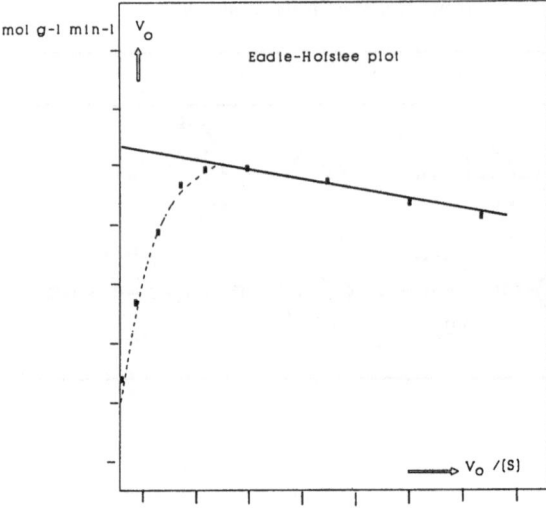

Fig. 27. An Eady-Hofstee plot showing a Km of 17.8 mM and V_{max} of 30.2 mol $gr^{-1}min^{-1}$ of the stereoselective enzymatic cyclisation of α-alkyl-*N*-carbamoyl-α-amino acids using hydantoinases

the aminonitrile to form the amino acid amide. Enzymatic hydrolysis results in the formation of the L-α-alkyl-α-amino acid and D-α-alkyl-α-amino acid amide.

The recycling of the unwanted isomer can for example be achieved by converting the amino acid into the starting material. The use of a relatively cheap whole cell biocatalyst contributes to the economic feasibility of the procedure. L-as well as D-amino acids can be prepared with a very high optical purity (approx. 100% e.e.). The biocatalyst is capable of stereoselectively hydrolyzing a broad range of structurally different α-alkyl-α-amino acid amides (see Fig. 25 and Table 20).

Table 21. Hydrolysis of 3,4-dimethoxy-α-methyl-phenylalanine amide by different strains (Ref. 131)

Example	Micro-organisms	Yield (%)	L-3,4-Dimethoxy-α-methylphenylalanine specific rotation $[\alpha]_d$ (C = 1, H_2O)*
1	*Bacillus subtilis* IFO-3026	74	−49°
2	*Canadida utilis* IFO-0396	36	−51°
3	*Rhizopus chinensis* IFO-4768	30	−47°
4	*Trichoderma viride* IFO-4847	28	−46°
5	*Nocardia asteroides* IFO-3424	49	−48°
6	*Mycobacterium smegmatis* NIHJ-1628	86	−52°
7	*Streptomyces griseus* IFO-3356	33	−49°
8	*Ustilago zeae* IFO-5346	66	−51°
9	*Pseudomonas fluorescens* IFO-3081	94	−53°
10	*Gibberella fujikuroi* IFO-5268	31	−47°
11	*Torulopsis candida* IFO-0768	17	−45°

Table 22. Effect of substrate concentration on the yield of the reaction

Concentration of the substrates (D,L-3,4-Dimethoxy-α-methylphenyl-alaninamide) (gew. %)	Yield on L-3,4-Dimethoxy-α-methyl-phenylalanine (%)
1	100
2	100
5	100
10	100
20	86
30	72
40	49

Some characteristics of the process are summarized below:

— Both L-and D-α-alkyl-α-amino acids can be produced.
— Permeabilized whole cells of *Mycobacterium neoaurum*, ATCC 25795 or crude enzyme preparations can be used.
— Very high stereoselectivity (>98% e.e.) and a remarkable broad substrate specificity (Table 20) is observed.
— The enzyme is active in a broad pH range; pH = 8.0–10.5.

The kinetics of the enzymatic hydrolysis were studied with D,α-methyl leucine amide as the substrate (see Fig. 26 and 27).

In addition to DSM the Ube company [131] has also reported an analogous biocatalytic route to α-methyl 3,4-dimethoxy-phenylalanine. The results in Table 21 illustrate that *Pseudomonas fluorescens* (IFO 3081) shows the highest conversion (94%) but the stereoselectivity is relatively low (93.4%). Concentrations up to 10% by weight could be used while still obtaining a 100% conversion (Table 22).

Hydantoinases

Watanabe et al. [132] at Kanegafuchi describe the stereoselective cyclization of N-carbamoyl-α-alkylated-α-amino acids into the corresponding D-hydantoins by using cultivated micro-organisms belonging to the genera *Aerobacter, Bacillus, Corynebacterium* etc. (see Fig. 28). D- and L-α-alkyl amino acids can be prepared from the D-hydantoins or the remaining L-N-carbamoyl α-alkylated amino acids, respectively.

For one example (see Table 23) the results with different micro-organisms are summarized.

Esterases

Another approach to synthesize (S)-α-methylphenylalanine, (S)-α-methyltyrosine and S-α-methyl-3,4-dihydroxyphenylalanine has been reported by Björkling et al. [133]. The enzyme used, catalyzes the hydrolysis of prochiral diesters. Pig liver esterase (see Fig. 29) catalyzed hydrolysis of compounds 1, 2 and 3 affording products (R)-4(45% e.e.), (R)-5 (82% e.e.) and R-6 (93% e.e.).

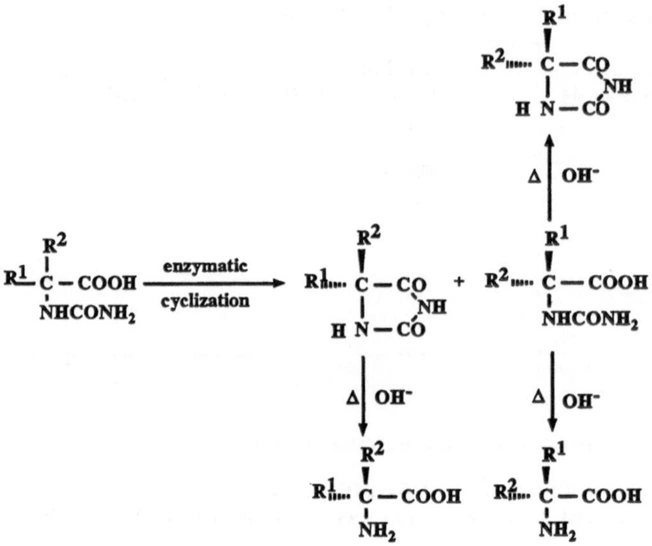

$$
R^2\text{''''}\!-\!\underset{\underset{H\,N-CO}{|}}{\overset{\overset{R^1}{|}}{C}}\!-\!\underset{NH}{CO}
$$

$\Delta \mid OH^-$

$$
R^1\!-\!\underset{\underset{NHCONH_2}{|}}{\overset{\overset{R^2}{|}}{C}}\!-\!COOH \xrightarrow[\text{cyclization}]{\text{enzymatic}} R^1\text{''''}\!-\!\underset{\underset{H\,N-CO}{|}}{\overset{\overset{R^2}{|}}{C}}\!-\!\underset{NH}{CO} \; + \; R^2\text{''''}\!-\!\underset{\underset{NHCONH_2}{|}}{\overset{\overset{R^1}{|}}{C}}\!-\!COOH
$$

$\Delta \mid OH^-$ $\Delta \mid OH^-$

$$
R^1\text{''''}\!-\!\underset{\underset{NH_2}{|}}{\overset{\overset{R^2}{|}}{C}}\!-\!COOH \qquad R^2\text{''''}\!-\!\underset{\underset{NH_2}{|}}{\overset{\overset{R^1}{|}}{C}}\!-\!COOH
$$

Fig. 28.

Table 23. Use of different hydantoinase strains in the ringclosure reaction of N-carbamoyl-α-methylamino acids (Ref. 132)

Micro-organism	Amount of produced 4-methyl-4-phenyl-imidazolidine-2,5'-dione (μmoles)	Amount of produced 6-fluoro-spiro [chroman-4,4'-imidazolidine-2',5' dione (μmoles)
Aerobacter cloacae IAM 1221	80	26
Agrobacterium rhizogenes IFO 13259	30	18 .
Bacillus badius IAM 11059	58	33
Bacillus species KNK 108 (FERM P-6056; FERM BP-887)	125	125
Bacillus stearothermophilus[a] IFO-12550	38	15
Coryne bacterium sepedonicum IFO 3306	104	76
Corynebacterium sepedonicum IFO 13259	77	51
Nocardia corallina IFO 3338	110	25

[a] Both cultivation and reaction was carried out at 50 °C

Table 24. Reaction conditions of the hydantoinase reaction and some physical constants of the products (Ref. 132)

(R-S)-N-carbamoyl α-methyl-α-amino acid	micro-org.	pH	T (°C)	t (hrs)	S-N-carbamoyl $[\alpha]_D^{25}$ (+92.0) (0.5, 0.1 N NaOH)
3,4-methylenedioxy-phenylalanine	bacillus species KNK108 FERMP-6056	7.2	37	48	R-hydantoin (+63.6) (c 0.5, 1 N NaOH)
Leucine	Corynebacterium sepedonicum IFO 13259			0	R-hydantoin (+0.8) (c 0.5, 0.1 N NaOH)
Phenylglycine	Nocardia corallina IFO 3338	7.2	37	20	S-carbamoyl (+8.0) (c 0.5, 0.1 N NaOH) R-hydantoin (−14) (c 0,5, 0.1 N NaOH) S-N-carbamoyl (+52) (c = 0.5, 0.1 N NaOH) S-hydantoin (+139.5) (c 0.5, 0.1 N NaOH)

Upon the addition of DMSO a large influence on the reaction rate and enantiomeric excess was observed when using α-chymotrypsin as the catalyst. The reaction rates with the substrates 2 and 3 were very low and incubation times of several weeks were needed.

Resolution of α-methylamino esters is also known. For example, Ananthara-maiah et al. [134] resolved D,L-α-methyltryptophan methylester, D,L-α-methyl-phenylalanine methylester and D,L-α-methylparafluorphenylalanine methylester using α-chymotrypsin at pH = 5.0. Non-enzymatic hydrolysis occurred simultaneously resulting in a low optical purity of the amino acids obtained (see Table 25).

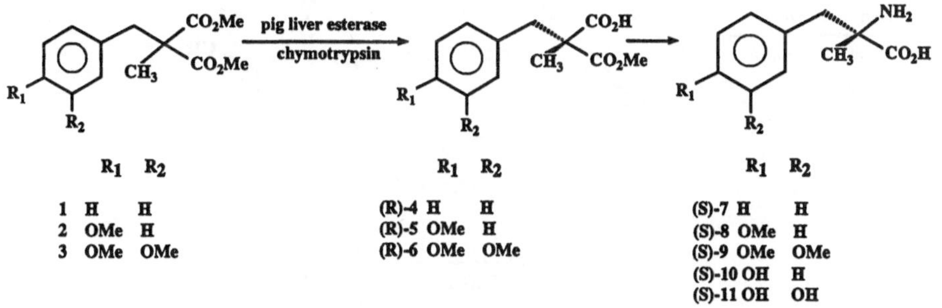

	R₁	R₂		R₁	R₂		R₁	R₂
1	H	H	(R)-4	H	H	(S)-7	H	H
2	OMe	H	(R)-5	OMe	H	(S)-8	OMe	H
3	OMe	OMe	(R)-6	OMe	OMe	(S)-9	OMe	OMe
						(S)-10	OH	H
						(S)-11	OH	OH

Fig. 29. PLE/Chymotrypsin hydrolysis of prochiral diesters

Table 25. Kinetic resolution of α-methyl amino acid esters (Ref. 134)

Product	Reaction time (h)	Yield %	m.p.	$[\alpha]_D^{25}$
D-α-MeTrp-OMe HCl	40	88	160° dec.	−28.3° (C9.9, MeOH)
L-α-MeTrp-OMe HCl	30	72	159° dec.	+27.8° (C9.9, MeOH)
D-α-MePhe-OMe HCl	35	79	112–115°	+23.6° (C9.8, N HCl)
L-α-MePhe-OMe HCl		66		
D-α-Me-p-F-Phe-Me HCl	35	78	143–144°	−11.2° (C1.0, MeOH)
D-α-Me-Phe				+19.3° (C1.0, H$_2$O)

Kinetic Resolution of α-Nitro-α-Methylcarboxylic Acid Esters

Wong and coworkers [135] have found that α-nitropropanoate esters are good substrates for enantioselective hydrolysis using chymotrypsin. Thus, by combining the rich chemistry of α-nitrocarboxylic acid esters with an enzymatic hydrolysis, these authors have developed a method to prepare α-nitropropanoic acid esters in high enantiomeric purity. These esters, potentially interesting themselves as amino acid analogs and as chiral starting materials, can easily be reduced to the corresponding α-methyl-α-amino acid esters using Adam's catalyst (PtO$_2$). The overall scheme outlined in Fig. 30 summarizes this approach for the preparation of optically active α-methyl-α-amino acids.

Enzymatic resolutions were carried out by using α-chymotrypsin, other enzymes like lipases and esterases were ineffective. Hydrolysis of the esters with lipases

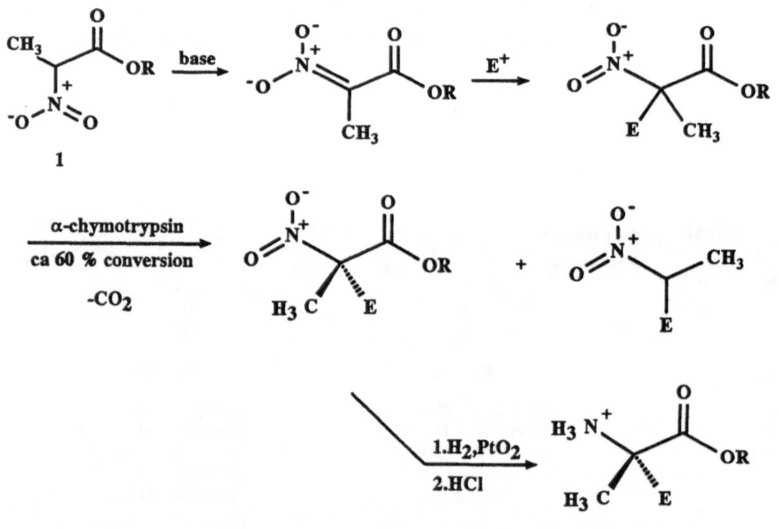

Fig. 30. Kinetic resolution of α-nitro-α-methyl carboxylic acids

Table 26. Kinetic resolution of α-methyl-α-nitro carboxylic acid esters (Ref. 135)

α-nitro carboxylic		enantiomeric excess (% e.e.)
acid ester	ester group	
13a	butyl	>95
13b	butyl	85
13c	methyl	39
13c	butyl	a
13d	butyl	75
13e	butyl	>95
13f	butyl	>95
13f	methyl	>95
13f	butyl	90
12	butyl	0

a: no hydrolysis

derived from *Candida cylindracea* (CCL) and from porcine pancreas (PPL) failed.
The enantiomeric excess of the product α-nitro esters kinetically resolved in this
way, were generally quite high, typically >95% (see Table 26)

Acylases

The hydrolytic action of Benzylpenicillin-acylase (BPA) from *Escherichia coli*
ATCC 9637 on a series of *N*-phenylacetyl α-methyl-α-amino acids containing an
aliphatic, aromatic or a polar side chain (*R*) on the chiral carbon atom, were
studied [136].

The results reported in Table 27 for compounds 14–20 show that as is found with
natural α-amino acids [137]-BPA hydrolysed the L-isomers more rapidly. However,
as can be seen by comparing the hydrolytic data for compounds 14–20 with those
of compounds 14a–20a the replacement of the α-hydrogen with a methyl group
greatly affects both the stereoselectivity and the rate of hydrolysis. With the signi-

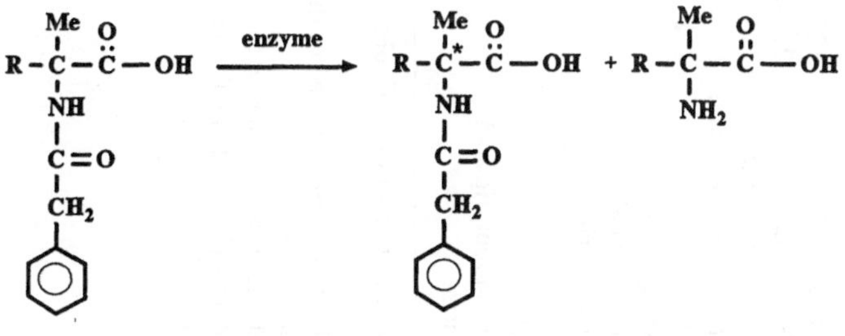

Fig. 31. Use of Benzylpenicillin acyclase in the resolution of *N*-Phenylacetyl-α-methyl-α-amino
acids

Table 27. Compounds examined in the hydrolysis of N-phenylacetyl-α-methylamino acids with benzylpenicillin acylase (Ref. 136)

Compound	M.p. of the starting racemic N-PA α-methyl-α-amino acid (°C)	Hydrolysis rate[a]	$[\alpha]_D$ of the amino acid released by the enzyme[b]	Obtained by HCl hydrolysis of the recoverred N-PA-derivative	Literature (L or D isomer)	Optical purity of the isolated α-methyl amino acid (%)
14. α-Methylphenylglycine	181–182	4.5	+ 74.9° (c = 2)[c]		+ 86.3° (L)	87
15. α-Methylglutamic acid	139–140	3.6	+ 9.0° (c = 5)[d]		+ 12.1° (L)	74
16. α-Methylaspartic acid	161–162	3.6	+ 35.0° (c = 2)		− 52.9° (D)	66
17. α-Methylserine	127–128	13.0	+ 4.0° (c = 2)		+ 6.3° (L)	63
18. α-Methylvaline	217–218	0.8		+ 2.3° (c = 2.2)	− 3.9° (L)	59
19. α-Methyl-α-amino-butyric acid (isovaline)	198–200	16.7		− 0.6° (c = 10)	+ 11.1° (L)	5
20. α-Methylleucine	182–183	7.6		− 2.5° (c = 5)	+ 34.0° (L)	7
14a. Phenylglycine		450	+112.0° (c = 1)	−113.0° (c = 1)	+113.0° (L)	
15a. Glutamic acid		360	+ 31.0° (c = 1)[g]	− 31.7° (c = 1)[g]	+ 31.5° (L)	
16a. Aspartic acid		360	+ 24.5° (c = 1)[g]	− 25.0° (c = 1)[g]	+ 24.9° (L)	
17a. Serine		270	− 7.5° (c = 1.2)	+ 7.8° (c = 1.5)	− 7.8° (L)	
18a. Valine		70	+ 6.0° (c = 2)	− 6.2° (c = 2)	+ 5.6° (L)	
19a. α-Aminobutyric acid		1400	+ 7.8° (c = 2)	− 8.0° (c = 2)	+ 7.9° (L)	
20a. Leucine		800	+ 15.5° (c = 1)[g]	− 15.7° (c = 1)[g]	+ 15.6° (L)	

[a] Expressed as micromoles of substrate hydrolyzed per h per mg of enzyme. 4.0 and 0.6 mg of BPA per mmol of substrate were used for compounds 14–20 and 14a–20a respectively.

[b] Optical rotations are referred to the amino acids obtained after one crystallization from H_2O-EtOH.

[c] Solvent 6 N HCl.

[d] The $[\alpha]_D$ reported is referred to L-α-methylvaline obtained by enzymatic resolution with CPA. Fractional crystallization (85% ethanol) of the brucine salt of (±) N-formyl-α-methylvaline gave optically pure L-α-methylvaline: $[\alpha]_D$-4.0 (c = 6, water), in accordance with the enzymatic resolution.

[g] Solvent 5 N HCl.

ficant exception of the *N*-phenylacetyl derivative of α-methylserine, whose corresponding α-hydrogen analog is not a good substrate for BPA (see Table 27) the rate of enzymatic hydrolysis is lowered by a factor of about 102 by the introduction of the methyl group on the α-carbon atom. This effect is analogous to, although more pronounced than, that recently found during the resolution of some unacylated α-methyl amino esters by chymotrypsin [134]. Data on optical purity reported in Table 27 clearly indicate the strong dependence of the stereoselectivity on the nature of the *R* substituent.

It is also known that one enzyme commonly used for such purposes, hog renal acylase I, is unable to catalyzes the hydrolysis of *N*-acetyl-α-methylphenylalanine [138]. Therefore, the use of commercially available bovine carboxypeptidase A (CPA) was studied ([139] see Fig. 32). *N*-trifluoracetyl derivatives of the amino acids were prepared because they are generally superior substrates for CPA when compared with the *N*-acetyl derivatives.

Both *N*-trifluoracetyl-α-methylphenylalanine and *N*-trifluoracetyl-α-methylvaline are digested stereospecifically by CPA, releasing the L-isomers and leaving the D-isomers intact. The protected derivatives can easily be separated from the amino acids by a simple extraction procedure. The absolute configuration of α-methylphenylalanine has been determined by the optical rotation of the free amino acid. This resolution procedure appears to be generally applicable to α-methyl amino acids whose side chains are compatible with the specificity of CPA.

Hog Kidney Amino Acylase

Keller et al. [140] reported on the preparation of the optical isomers of 2-trifluoromethylalanine by partial hydrolysis of the racemic *N*-trifluoroacetyl derivative with hog kidney aminoacylase (HKA) (EC 3.5.1.14, see Fig. 33). Racemic 2-trifluoromethylalanine is a powerful irreversible inhibitor of *Pseudomonas cepacia* 2,2-dialkylglycine decarboxylase (EC 4.1.1.64).

Thus, hog kidney aminoacylase catalyzes the hydrolysis of *N*-trifluoroacetyl-*R*(+)2-trifluoromethylalanine with 99.1 % enantioselectivity.

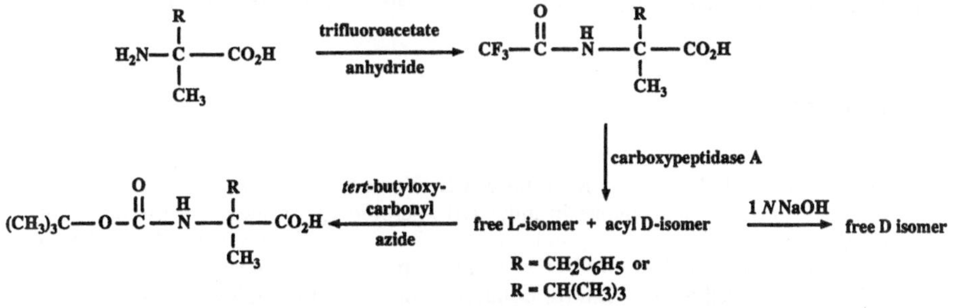

Fig. 32. Acylase catalyzed hydrolysis of *N*-trifluoro-acetylated α-alkyl amino acids

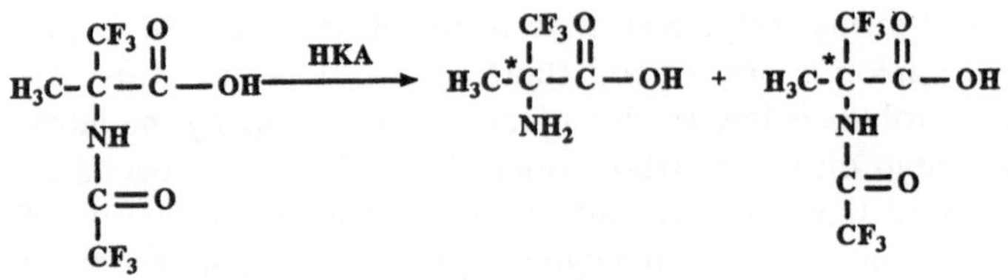

Fig. 33. Resolution of α-trifluoro-methylalanine

6 Conclusions and Future Prospects

The commercial interest in enantiomerically pure amino acids has triggered a worldwide research effort towards the synthesis of this important class of compounds, which is reflected in a tremendous number of papers and patent applications published in the last decade. However, the introduction of successful novel commercial processes has been rather limited. This is mainly due to the strength of the established 'classical' procedures like the acylase process, purely chemical methods or, for the production of natural amino acids, microbial production techniques. Evidently, these established processes have been greatly improved by implementing the latest scientific and technological developments.

Still, a number of novel chemo-enzymatic processes have been commercialized recently and it is to be expected that in the future an ever increasing number of enantiomerically pure (synthetic) amino acids will be produced both by the application of established procedures and by newly developed chemo-enzymatic methods. Highly promising developments are the discovery of new enzyme activities like D-specific acylases and D-specific aminopeptidases, which, if used in combination with the appropriate racemases, will dramatically improve the kinetic resolution processes in use today. Moreover, it is to be expected that goal-orientated approaches, like in vivo protein engineering, will be increasingly applied for the selective generation of enzyme activities with new and improved properties.

Acknowledgements: Special thanks are due to Mr. W. Kortenoeven for his exceptionally valuable assistance in the preparation of the manuscript.

7 References

1. Aida K, Chibata I, Nakayama K, Takinami K, Yamada H (eds) (1986) Biotechnology of amino acid production: progress in industrial microbiology, Elsevier, Amsterdam, vol 24
2. Ariëns EJ (1983) Stereoselectivity in bioactive agents: general aspects. In: Ariëns EJ, Soudijn W, Timmermans, PBMWM (eds) Stereochemistry and biological activity of drugs, Blackwell, Blackwell Sc. Pb. Oxford, London, Edinburgh, Boston, Melbourne, p 11
3. Ariëns EJ, Van Rensen JJS, Welling W (eds) (1988) Stereoselectivity of pesticides, Elsevier, Amsterdam

4. Yonaha K, Soda K (1986) Application of stereoselectivity of enzymes: synthesis of optically active amino acids and α-hydroxy-acids, and stereospecific isotope-labeling of amino acids, amines and coenzymes. In: Fiechter A (ed) Advances in biochemical engineering/biotechnology, Springer, Berlin Heidelberg New York vol 33, p 95
5. Williams RM (1989) Synthesis of optically active amino acids, Pergamon, Oxford
6. Ajinomoto, French Patent 787,199 (1977)
7. Kanegafuchi, German Patent Application, 704,212 (1976)
8 a. Chibata I, Tosa T, Sato T (1988) Application of immobilized biocatalysts in pharmaceutical and chemical industries. In: Rehm HJ, Reed G (eds), Biotechnology vol 7a (enzyme technology), VCH Verlag, Weinheim, chap. 12b, p 653
  b. Chibata I, Tosa T, Sato T, Sato T (Acad. Press. New York) In: Mosbach (ed) Methods in enzymology, Academic, New York, vol 44, p 746
9. Kanegafuchi, German Patent 2,704,212 (1977)
10. Leuchtenberger W, Karrenbauer M, Plöcker U (1984) Forum Mikrobiologie, 7: 40
11. Chibata I, Tosa T (1981) Ann. Rev. Biophys. Bio-engin. 10: 197
12. Wandrey C, Flaschel E (1979) Process developments and economical aspects in enzyme engineering. Acylase-L-methionine system. In: Ghose TK, Fiechter A, Blakeborough N (eds) Advances in biochemical engineering, Springer, Berlin Heidelberg New York, vol 12, p 147
13. Wandrey C, Wichmann R (1983) Produktion von L-aminosäuren im membran-reaktor, Jahresbericht 1982/1983 der Kernforschungsanlage Jülich GmbH
14. Chibata I (1974) Optical resolution of DL amino acids. In: Kanedo T, Izumi Y, Chibata I, Itoh T (eds) Synthetic production and utilization of amino acids, Wiley, New York, p 33
15. Chenault HK, Dahmer J, Whitesides GM (1989) J. Am. Chem. Soc. 111: 6354
16. Marschall R, Birnbaum SM, Greenstein JP (1956) J. Am. Chem. Soc. 78: 4636
17. Gilles I, Löffler H-G, Schneider FZ (1981) Naturforsch. C. 36C: 715
18. Hong-Yong Cho, Tanizawa K, Tanaka H, Soda K (1987) Agric. Biol. Chem. 51: 2793
19. Maestracci M, Bui K, Thiery A, Arnaud A, Galzy P (1988) The amidases from a *Brevibacterium* strain: study and applications. In: Fiechter A (ed) Advances in biochemical engineering/biotechnology, Springer, Berlin Heidelberg New York, vol 36, p 69
20. Sheldon RA, Schoemaker HE, Kamphuis J, Boesten WHJ, Meijer EM (1988) Enzymatic methods for the industrial synthesis of optically active compounds. In: Ariëns EJ, van Rensen JJS, Welling W (eds) Stereoselectivity of pesticides, Elsevier, Amsterdam, p 409
21. Kamphuis J, Kloosterman M, Schoemaker HE, Boesten WHJ, Meijer EM (1987) Chiral intermediates and applications. In: Neijssel OM, Van der Meer RR, Luyben KChAM (eds) Proceedings 4th European Congress on Biotechnology. Elsevier, Amsterdam. vol 4, p 331
22. Kamphuis J, Hermes HFM, van Balken JAM, Schoemaker HE, Boesten WHJ, Meijer EM (1990) Chemo-enzymatic synthesis of enantiomerically pure α-H and α-alkyl-α-amino acids and derivatives In: Lubeck G, Rosenthal AE (eds) Amino acids, synthesis, biology and medicine, ESCOM Science publishers, Leiden, p 119
23. Meijer EM, Boesten WHJ, Schoemaker HE, van Balken, JAM (1985) Use of biocatalysts in the industrial production of specialty chemicals. In: Tramper J, van der Plas HC, Linko P (eds) Biocatalysts in Organic Syntheses, Elsevier, Amsterdam, p 135
24. Boesten WHJ, Dassen BHN, Kerkhoffs PL, Roberts MJA, Cals MJH, Peters PJH, van Balken JAM, Meijer EM, Schoemaker HE (1986) Efficient enzymatic production of enantionerically pure amino acids. In: Schneider MP (ed) Enzymes as catalysts in organic syntheses, NATO ASI Series C, Reidel, Dordrecht, vol 178, p 355
25. DSM/Stamicarbon, British Patent 1,548,032 (1976)
25. DSM/Stamicarbon, U.S. Patent 3,971,000 (1976)
27. Vriesema BK, ten Hoeve W, Wijnberg H, Kellogg RM, Boesten WHJ, Meijer EM, Schoemaker HE (1986) Tetrahedron Letters 26: 2045
28. NOVO-DSM/Stamicarbon U.S. Patent 4,080,259 (1978)
29. DSM/Stamicarbon, U.S. Patent 4,172,846 (1979)
30. DSM/Stamicarbon, Dutch Patent Application 8,403,093 (1984)
31. DSM/Stamicarbon, Dutch Patent Application 8,501,093 (1985)

32. Shadid B, van der Plas HC, Boesten WHJ, Kamphuis J, Meijer EM, Schoemaker HE (1990) Tetrahedron 46: 913
33. Duke CC, Macleod JK, Summons RE, Letham DS, Parker CW (1978) Aust. J. Chem. 31: 1291
34. Feenstra RW, Stokkingreef EHM, Reichwein AM Lousberg WBH, Ottenheijm HCJ, Kamphuis J, Boesten WHJ, Schoemaker HE, Meijer EM (1990) Tetrahedron 46: 1745
35. Maehr H (1971) Pure Appl. Chem. 28: 603
36. Keller-Schierlein W, Prelog V, Zähner H (1964) Forschr. Chem. Org. Naturstoffe, 22: 279
37. Takahashi J, Ohashi T, Kii Y, Kumagai H, Yamada H (1979) J. Ferment. Technol. 57: 328
38. Ajinomoto, French Patent 7,817,199 (1977)
39. Guivarch M, Gillonier G, Brunie JC (1980) Bull. Soc. Chim. Fr. 91
40. Cecere F, Galli G, Morisi F (1975) FEBS Lett. 57: 192
41. Snamprogetti, British Patent 1,506,067 (1978)
42. Dinelli D, Marconi W, Cecere F, Galli G, Morisi F (1978) A new method for the production of optically active amino acids. In: Pye EK, Weetall HW (eds) Enzyme engineering, Plenum, New York, vol 5, p 477
43. Kanegafuchi, Britisch Patent 1,572,316 (1978)
44. Olivieri R, Fascetti E, Angelini L. Degen L (1981) Biotechnol. Bioengin. 23: 2173
45. Sambale C, Kula M-R (1988) J. of Biotechn. 7: 49
46. Kameda Y, Toyoura E, Kimura Y (1958), Nature (London) 181: 1295
47. Kameda Y, Toyoura E, Yamazoe M, Kimura Y, Yasuda Y (1952), Nature, 170: 888
48. Fukagawa Y, Kubo K, Ishikura T, Kouno K (1980) J. Antibiot. 33: 543
49. Kubo K, Ishikura T, Fukagawa Y (1980) J. Antibiot. 33: 550
50. Kubo K, Ishikura T, Fukagawa Y (1980) J. Antibiot. 33: 556
51. Sugi M, Suzuki H (1978) Agric. Biol. Chem. 42: 107
52. Sugi M, Suzuki H (1980) Agric. Biol. Chem. 44: 1089
53. Ying-Chieh, T, Ching-Ping T, Kuang-Ming T, Hsiao H, Ling-Yun C (1988) Appl. and Environm. Microbiol. 984–989
54. Asano Y, Nakazawa A, Kato Y, Kondo K (1989) Angew. Chem. 4: 450
55. Kato Y, Asano Y, Nakazawa A, Kondo K (1989) Tetrahedron 45: 5743
56. Asano Y, Mori T, Hanamoto Y, Kato Y, Nakazawa A (1989) Biochem. Biophys. Res. Comm. 1: 470
57. Syldatk C, Cotoras D, Möller A, Wagner F (1986) Biotech. forum 3: 10
58. Groß C, Syldatk C, Mackowiak V, Wagner F (1990) J. of Biotechn. (in press)
59. Yokozeki K, Sano K, Eguchi C, Iwagami H, Mitsugi K (1987) Agric. Biol. Chem. 51: 729
60. Yokozeki K, Hirose Y, Kubota K (1987) Agric. Biol. Chem. 51: 737
61. Ajinomoto, Japanese patent application 095674 (1975); Ajinomoto, Japanese patent application 110124 (1976); Tanabe, Japanese patent 031514 (1983); Schering AG, German patent application DE 702384 (1988)
62. Tanabe, Japanese patent 132801 (1984)
63. Tanabe, Japanese patent 108380 (1987)
64. Schering AG, German patent DE 703384 (1987)
65. Ajinomoto, Japanese patent, JP 167535 (1986)
66. Groß C, Syldatk C, Wagner F (1987) Development of a process for the production of aromatic L-amino acids from D,L-5-monosubstituted hydantoins. In Neijssel OM, van der Meer RR, Luyben KChAM (eds), Proceedings 4th European Congress on Biotechnology, vol 2, Elsevier, Amsterdam p 249, and references cited
67. Inagaki K, Tanizawa K, Tanaka H, Soda K (1987) Agric. Biol. Chem. 51: 173
68. Neidhart DJ, Distefano MD, Tanizawa K, Soda K, Walsh CT, Petsko GA (1987) J. Biol. Chem. 262: 15323
69. Takeda Chem Ind, European patent application 0,304,021 (1988)
70. Rütgerswerke DE 37153 (1988)
71. Cheetham PSJ (1987) Enzyme Microb. Technol. 9: 194
72. Fukumura T (1976) Agric. Biol. Chem. 40: 1687
73. Fukumura T (1976) Agric. Biol. Chem. 40: 1695
74. Fukumura T (1977) Agric. Biol. Chem. 41: 1327

75. Fogarty WM, Kelley C (1980) Amylases, amyloglucosidases and related glucanases. In: Rose AH (ed) Microbial enzymes and bioconversions, Academic, New York, p 115
76. Rowlands RT (1984) Enzyme Microb. Technol. 6: 289
77. Groß C, Syldatk C, Wagner F (1987) Biotechnol. Techniques, 1: 85
78. Morin A, Hummel K, Kula M-R (1986) Biotechnol. Lett., 8: 573
79. Winter G, Fersht AR (1984) Trends in Biotechnology, 2: 115
80. Wells AJ, Estell DA (1988) Trends in biochem. sciences 13: 291
81. Harder W, Kuenen JG (1977) J. Appl. Bacteriol. 43: 1
82. Dijkhuizen D, Hartl DL (1983) Micr. Reviews, 47: 150
83. Harder W (1987) Microbial physiology and biotechnological innovation. In: Alberghina L, Frontali L, Hamer G (eds) Dechema monographs, VCH Verlag, Weinheim, Vol 105, p 1
84. Clarke PH (1987) Experimental enzyme evolution and the design of novel biocatalysts. In: Alberghina L, Frontali L, Hamer G (eds) Dechema monographs, VCH Verlag, Weinheim, vol 105, p 13
85. Mortlock RP (1986) Trends in biotechnology, 65
86. Beacham IR (1987) FEMS Microbiology Reviews, 46: 409
87. Mortlock, RP (ed) (1984) Microorganisms as model systems for studying evolution, Plenum, New York
88. Hartley BS (1966, May) Adv. Sci. 47
89. Liao H, MacKenzie T, Hagerman R (1986) Proc. Nat. Acad. Sci. USA, 83: 576
90. Sonnleitner B (1983) Biotechnology of thermophilic bacteria-growth, products and application. In: Fiechter A (ed) Advances in biochemical engineering/biotechnology, Springer, Berlin Heidelberg New York, Vol 28, p 69
91. Daniel RM, Cowan DA, Morgan HW, Curran P (1982) Biochem. J. 207: 641
92. DSM/Stamicarbon-NOVO/Nordisk, European patent application 0307023 (1989)
93. Hughes EJ, Shapiro MK, Houghton JE, Ornston LN (1988) J. Gen. Microbiol. 134: 2877
94. Chibata I, Tosa T and Sato T (1986) Aspartic acid. In: Aida K, Chibata I, Nakayama K, Takinami K, Yamada H (eds) Biotechnology of amino acid production: progress in industrial microbiology, Elsevier, Amsterdam, vol 24, p 144
95. Chibata I, Tosa T and Sato T (1985) Immobilized biocatalysts to produce amino acids and other organic compounds. In: Enzymes and immobilized cells in biotechnology. Benjamin/Cummings p 37
96. Leuchtenberger W, Plöcker U (1988) Chem. Ing. Techn. 60: 16
97. Hummel W, Kula M-R (1989) Eur. J. Biochem. 184: 1
98. Calton GJ (1987) The enzymatic production of amino acids. In: Neijssel OM, Van der Meer RR, Luyben KChAM (eds) Proceedings 4th European Congress on Biotechnology. Elsevier, Amsterdam. vol 4, p 693
99. Nakamichi K, Nishida Y, Nabe K, Tosa T (1984) Applied Biochim. Biotechn. 11: 367
100. Hamilton BK, Hsiao H, Swann WE, Anderson DM, Delente JJ (1985) Trends in Biotechn. 3: 64
101. Evans CT, Hanna K, Payne C, Conrad D, Misawa M (1987) Enzyme Microb. Techn. 9: 417
102. De Boer L, Dijkhuizen L (1990) Microbial and enzymatic processes for L-phenylalanine production. In: Fiechter A (ed) Advances in biochemical engineering/biotechnology, Springer Berlin Heidelberg New York (in press)
103. Tanabe, Japanese patent 22881 (1967)
104. Tanabe, US Patent 3,871,959 (1975)
105. Mitsui Toatsu, Japanese patent 212292 (1986)
106. Acad. of Science Tatzhik, Russian Patent SU 403669 (1973)
107. Ajinomoto, British Patent 1,345,113 (1974) Ajinomoto
108. Yamada S, Maeshima H, Wada M, Chibata I (1973) Appl. Microbiol. 25: 636
109. Dalaney FL (1967) Microbial production of amino acids. In: Pepple HL (ed) Microbial Technology, Reinhold, New York, p 308
110. Meister A (1965) Biochemistry of amino acids, Academic, New York, vol 1, p 113
111. Yamada S, Wada M, Izuo N, Chibata I (1976) Applied Environm. Microbiology, 32: 1
112. Kyoto University, Japanese patent 205740 (1987)
113. Hoechst AG German patent DE 3733506 (1989)

114. Kise H, Shirato H, Nositomo H (1987) Bull. Chem. Soc. Jpn. 60: 3613
115. Ajinomoto Japanese patent 80/88697 (1980); Mitsui Toatsu Japanese patent 86/177992 (1986)
116. Toray Japanese patent 029709 (1987); Kyowa Hakko, Hakko to Taisha, 15, 89–94 (1967)
117. Pandey RC, Meng H, Cook JC, Rinehart KL (1977) J. Am. Chem. Soc. 99: 5202
118. Fox RO, Richard FM (1982) Nature 300: 325
119. Goodman M (1985) Biopolymers 24: 137
120. Marshall OR (1982) in 'Chemical Regulation of Biological Mechanisms; Cubitt AG and Creighton AM (Eds), The Chemical Society, London
121. Barone V, Fraternali F, Cristinziano PL, Lelj F and Rosey, A (1988) Biopolymers 27: 1673 and references cited therein
122. Toniolo C, Benedetti E (1988) ISI Atlas of Science Biochemistry, 225–230
123. Kruizinga WH, Bolster J, Kellogg RM, Kamphuis J, Boesten WHJ, Meijer EM, Schoemaker HE (1988) J. Org. Chem. 53: 1826
124. Schöllkopf U (1983) Topics in Current Chem. 109: 65
125. Beck AK, Seebach D (1988) Chimia, 42: 142 and references cited therein
126. Fadel A, Salaun J (1987) Tetrahedron Lett. 28: 2243
127. Schöllkopf U, Tolle R, Egert E, Nieger M (1987) Liebigs Ann. Chem. 399
128. DSM/Stamicarbon, Eur. patent 0.150.854 (1984)
129. DSM/Stamicarbon Eur. patents 0.231.548, 0.236.591, 0.232.562 (1986)
130. Duchateau A, Crombach M, Kamphuis J, Boesten WHJ, Schoemaker HE, Meijer EM (1989) J. of Chromatography 471: 263
131. Ube, German patent DE 3217908 (1989)
132. Kanegafuchi, European patent application 0.175.312 (1985)
133. Björkling F, Boutelje J, Gatenbeck S, Hult K, Norin T (1985) Tetrahedron Lett 40: 4957
134. Anantharamaiah EM, Zoeske RW (1982) Tetrahedron Lett. 23: 3335
135. Lalonde JJ, Bergbreiter DE, Wong CH (1988) J. Org. Chem. 53: 2323
136. Rossi D, Calcagni A (1985) Experienta, 41: 35
137. Rossi D, Romeo A, Lucente G (1978) J. Org. Chem. 43: 2576
138. Rossi D, Lucente G, Romeo A (1977) Experientia, 33: 1557
139. Turk J, Panse GT, Marshall GR (1975) J. Org. Chem. 40: 953
140. Keller JW, Hamilton BJ (1986) Tetrahedron Lett. 27: 1249

Biochemistry and Biotechnology of Amino Acid Dehydrogenases

Toshihisa Ohshima[1] and Kenji Soda[2]

[1] Department of Chemistry, Kyoto University of Education, Fushimi-ku, Kyoto 612, Japan
[2] Laboratory of Microbial Biochemistry, Institute for Chemical Research, Kyoto University, Uji, Kyoto-Fu 611, Japan

Over the last decade, amino acid dehydrogenases such as alanine dehydrogenase (Ala DH), leucine dehydrogenase (Leu DH), and phenylalanine dehydrogenase (Phe DH) have been applied to the enantiomer-specific synthesis and analysis of various amino acids. In particular, amino acid dehydrogenases from thermophiles have received much attention because of their high stability. Their productivity was enhanced and the purification facilitated by the gene cloning. The advances in biotechnological applications of these enzymes are based on fundamental studies concerning characteristics of the enzymes and reaction mechanism as described in this chapter. Further elucidation of the structure and function of these enzymes based on genetic engineering and protein engineering may enable their properties to be improved for their future uses in biotechnology.

Advances in Biochemical Engineering/
Biotechnology, Vol. 42
Managing Editor: A. Fiechter
© Springer-Verlag Berlin Heidelberg 1990

1 Introduction

Amino acid dehydrogenase (EC 1.4.1.—) catalyze the reversible deamination of
amino acids to the corresponding keto acids in the presence of pyridine nucleotide
coenzymes, NAD(P):

$$\underset{\underset{\text{COOH}}{|}}{\overset{\overset{\text{R}}{|}}{\text{H}-\text{C}-\text{NH}_2}} + \text{NAD(P)} + \text{H}_2\text{O} \rightleftharpoons \underset{\underset{\text{COOH}}{|}}{\overset{\overset{\text{R}}{|}}{\text{C}=\text{O}}} + \text{NH}_3 + \text{NAD(P)H} + \text{H}^+$$

The dehydrogenation is accompanied by deamination, and in this respect amino
acid dehydrogenases are different from many other NAD(P)-dependent dehydrogenases like lactate dehydrogenase and alcohol dehydrogenase which act on a
hydroxyl group of the substrates. More than ten kinds of amino acid dehydrogenases have so far been found in various organisms (Table 1). Amino acid dehydrogenases except *meso*-diaminopimelate D-dehydrogenase (EC 1.4.1.16) act
on the L-form of amino acids as the substrate. Although glutamate dehydrogenase
(Glu DH) is found ubiquitously in various organisms including animals, insects,
plants, and microorganisms, except in the majority of the bacilli, the other amino
acid dehydrogenases occur in microorganisms and/or plants.

Table 1. NAD(P)-dependent amino acid dehydrogenases

Enzyme[a]	EC number	Coenzyme	Enzyme[a]	EC number	Coenzyme
Ala DH	1.4.1.1	NAD	3,5-Diamino-		
Glu DH	1.4.1.2	NAD	hexanoate DH	1.4.1.11	NAD
Glu DH	1.4.1.3	NAD(P)	2,4-Diamino-		
Glu DH	1.4.1.4	NAD	pentanoate DH	1.4.1.12	NAD(P)
Ser DH	1.4.1.7	NAD	Lys DH	1.4.1.15	NAD
Val DH	1.4.1.8	NAD, NADP	Diamino-		
			pimelate DH	1.4.1.16	NADP
Leu DH	1.4.1.9	NAD	Phe DH	1.4.1.—	NAD
Gly DH	1.4.1:10	NAD	Try DH	1.4.1.—	NAD(P)

[a] DH: dehydrogenase

Amino acid dehydrogenases provide a route for interconversion of inorganic
nitrogen with organic nitrogen, and, in other words, serve as a connecting link
between amino acid and organic acid metabolism. The equilibrium for the reactions is greatly in favor of amino acid formation. The value of the equilibrium
constant, K, is as follows:

$$K = [\text{keto acid}]\,[\text{NH}_3]\,[\text{NAD(P)H}]\,[\text{H}^+]/[\text{amino acid}]\,[\text{NAD(P)}]\,[\text{H}_2\text{O}]$$

The K values of the reactions of glutamate dehydrogenase, alanine dehydrogenase
(Ala DH), leucine dehydrogenase (Leu DH), and phenylalanine dehydrogenase

(Phe DH) are in the region of 10^{-15} M [1–4]. However, the metabolic functions of amino acid dehydrogenases differ with respect to the individual enzymes and organisms. For example, *Neurospora crassa* produces two distinct Glu DHs; one is specific for NADP and has a biosynthetic role, and the other one requires NAD and functions biodegradatively [5, 6]. Because of their extremely small equilibrium constants described above the enzymes are useful as catalysts for the production of amino acids from keto analogs. NAD(P)H absorbs at 340 nm, and the enzyme activity is easily and accurately assayed spectrophotometrically. Therefore, amino acid dehydrogenases are applicable to the determination of substrates such as amino acids and keto acids. In this chapter we describe the biochemical and biotechnological aspects of amino acid dehydrogenases.

2 Molecular and Catalytic Properties of Amino Acid Dehydrogenases

2.1 Glutamate Dehydrogenases

Glu DHs (EC 1.4.1.2–4) catalyze the oxidative deamination of L-glutamate to α-ketoglutarate with concomitant reduction of NAD(P). The enzymes play important physiological roles between the metabolism of glutamate, which represents a point of convergence in amino acid metabolism, and α-ketoglutarate, one of the important constituents of the citric acid cycle. Most of the enzymes are classified into two metabolic categories [7]. One category generally NADP-specific) is involved in ammonia assimilation (NADP-Glu DH, EC 1.4.1.4), and other category (generally NAD-specific) participates (NAD-Glu DH, EC 1.4.1.2) in glutamate catabolism. This strict coenzyme specificity is not found in vertebrate Glu DH, (NAD(P)-Glu DH, EC 1.4.1.3), which utilizes both NAD and NADP with a similar efficiency (Table 2), and occurs chiefly in the matrix of the mitochondria [26]. Either NADP or NAD-Glu DHs are found in higher plants and microorganisms except filamentous fungi such as *Neurospora crassa* and yeasts such as *Candida* and *Saccharomyces*. Although plant NAD-GDH DH is in the mitochondria, the NADP-enzyme is located in the chloroplasts [7]. NADP-Glu DH and NAD-Glu DH in many fungi and yeasts are present in the mitochondria and the cytosol, respectively. In contrast, NAD(P)-Glu DH occurs chiefly in the matrix of the mitochondria in vertebrate cells although it is synthesized in the microsomes [7].

Most Glu DHs consist of six identical subunits with a molecular weight between 42,000 and 63,000 (Table 2). The protomers have no catalytic activity. The smallest unit of enzymatically active mammalian NAD(P)-Glu DH (EC 1.4.1.3) is the hexamer with a molecular weight of about 330,000. At high concentration of the enzyme (above 0.1 mg/ml), the hexamers are known to undergo reversible unlimited aggregation up to a molecular weight of about 2,000,000. The aggregates show electron microscopically a rod-like structure with a helical array [27]. Besides the enzyme concentration, the allosteric modifiers of the enzyme

T. Ohshima and K. Soda

Table 2. Major sources and molecular properties of purified Glu DHs

Sources	Coenzyme	M_r (10^3) (subunits)	Crystalline form	Ref.
(Animal)				
Human liver	NAD(P)	300 (6 × 50,000)	Needles	[8]
Bovine liver	NAD(P)	330 (6 × 55,393)	Needles	[7]
Porcine liver	NAD(P)	310 (6 × 52,000)	Needles	[7]
Chicken liver	NAD(P)	306–346 (6 × 55,658)	Thin hexagonal plates	[7]
Rat liver	NAD(P)	330–370		[7]
Frog liver	NAD(P)	250	Stubby needles	[7]
Dogfish liver	NAD(P)	310–350 (6 × 51–56,000)	Needles	[7]
Mealworm fat body	NAD(P)	340 (6 × 57,000)		[9]
(Plant)				
Pisum sativum (root)	NAD	210 (—)		[7]
Turnip (mitocondria)	NAD	— (43,000)		[10]
Sphaerostilbe repens	NADP	280 (6 × 48,000)		[11]
(Microorganisms)				
Euglena gracilis	NADP	180 (4 × 45,000)		[12]
Chlorella sarokiniana	NADP	290–410 (6 × 58,000)		[7]
	NAD	180 (4 × 45,000)		[7]
Ankistrodesmus braunii	NAD(P)	380 (6 × 51,900)		[13]
Neurospora crassa	NAD	480 (4 × 116,000)	Trigonal	[7]
	NADP	288 (6 × 48,438)		[7]
Aspergillus nidulans	NAD	(– × 110,000)		[14]
Candida utilis	NAD	460 (4 × 116,000)		[7]
Saccharomyces	NAD	450 (4 × 100,000)		[15]
cerevisea	NADP	270–290 (6 × 48,000)		[7]
Phormidium laminosum	NADP	280 (—)		[16]
Salmonella typhimurium	NADP	280 (—)		[7]
Escherichia coli	NADP	245 (6 × 46,000)		[7]
Pseudomonas aeruginosa	NADP	280 (6 × 45,000)		[17]
Nitobacter				
hamburgensis	NADP	310 (6 × 48,000)		[18]
Bacteroides fragilis	NADP	290 (6 × 49,000)		[19]
Bacillus subtilis	NAD(P)	250–270 (4 or 6 × 57,000)		[20]
B. stearothermophilus	NAD	240 (6 × 40,000)		[21]
Clostridium botulinum	NAD	250.8 (6 × 42,500)		[22]
Clostridium SB4	NAD	275	Rhombohedoric	[7]
Thiobacillus novellus	NAD	110–113		[7]
	NADP	120–140		[7]
Lactobacillus fermentum	NADP	300 (6 × 50,000)		[23]
Mycoplasma laidawii	NAD(P)	240–260 (6 × 48,000)		[7]
Hyphomicrobum	NADP	380 (6 × 63,400)		[7]
Methylophilus				
metanolovorus	NADP	300 (6 × 49,000)		[24]
Halobacterium halobium	NAD	148		[25]
	NADP	215		[25]

such as GTP and ADP, hormones and drugs also affect the degree of the reversible aggregation. Some physiological significance of this aggregation with respect to the regulation of the enzyme was suggested, but has not been clearly elucidated [7, 28]. The reversible aggregation does not occur with the enzymes from rat liver, digfish liver, tuna liver and non-animal sources [7]. The fungal and yeast NAD-Glu DHs consist of 4 subunits (molecular weight of 100,000 to 120,000) and their production is induced by glutamate [14, 15, 29, 30]. The *Candida* and *Saccharomyces* NAD-Glu DHs are regulated by a reversible phosphorylation/dephosphorylation mechanism [29, 30]. The corresponding NADP-Glu DHs are hexameric, and have a subunit molecular weight of 48,000; they are induced by inorganic nitrogen which is used metabolically for glutamate formation and they are rapidly degraded in response to nitrogen or carbon limitation [31]. The substrate specificity of Glu DH is rather high (Table 3). L-Amino acids other than L-glutamate and keto acids other than α-ketoglutarate are utilized very slowly.

Table 3. Substrate specificity of bovine liver [32], frog liver [33], and *Nitorosomonas europaes* [34] Glu DHs

Substrates	Relative activity (pH 9.0 and 8.7*)		
	Bovine liver	Frog liver	*N. europaea**
L-Glutamate	100	100	100
L-Homocysteinesulfinate	65	—	—
L-α-Aminobutyrate	6.4	38	0
L-Norvaline	5.8	23	0
L-Leucine	1.2	12	0
L-Valine	0.68	16	0
DL-Norleucine	0.42	1.7	—
L-Isoleucine	0.5	21	0
L-Alanine	1.0	4.3	0
L-Ornithine	0	22	—
L-Aspartate	0	4.3	0
L-Lysine	0	6.8	—
L-Glutamine	6.2	32	—
D-Glutamate	0	0	0

In particular, the enzymes from microorganisms such as *N. europaea* [34] act specifically on L-glutamate and α-ketoglutarate. Ammonia is the best amino donor in the amination, and the amide moiety of glutamine and asparagine are also used as the amino donor. The K_m values for various Glu DHs are summarized in Table 4. The values for the substrates, in particular ammonia and coenzymes, vary rather widely with the source of the enzymes. The optimum pHs for oxidative deamination and reductive amination are at the alkaline pH range.

2.2 Alanine Dehydrogenase

Ala DH catalyzes the reversible deamination of L-alanine to pyruvate. The enzyme occurs in both prokaryotic and eukaryotic microorganisms (Table 5). The enzyme of *Bacillus* species is induced by DL-alanine in the medium [35, 36], and

Table 4. Optimum pH and Michaelis constants for substrates of Glu DH reactions

Sources	Optimum pH		K_m values (mM)						
	Deamination	Amination	L-Glu	2-OGA	NH₃	NAD	NADH	NADP	NADPH
Bovine liver	8.5–9.0	7.8	1.8	0.7	3.2	0.7	0.024	0.047	0.025
Frog liver	8.5–9.0	–	1.8	5.0	0.5	–	0.20	0.50	0.20
Pea root	8.2	7.9	7.3	3.3	38	0.65	0.86	0.038	0.13
Chlorella	9.2	7.2	32	12	68				
Neurospora crassa (NADP)	8.6–9.0	7.6	4.5	5.3	10			0.050	0.125
(NAD)	–	–	5.5	4.6	17	0.33	0.55		
Clostridium SB₄	9.4	7.8	1.8	0.65	0.32	0.010	0.010		
Nitosomas europaea	8.7	7.7	6.7	4.3	16			0.0079	0.049
Halobacterium halobium	9.2	8.0	4.0	20.2	450	0.30	0.07		

Table 5. Molecular properties of Ala DHs

Source	M_r (10^3) (subunit structure)	Degree of purification	K_m values (mM)					Ref.
			L-Ala	NAD	Pyr	NH_3	NADH	
Bacillus subtilis	228 (6 × 38,000) (needle-like crystal)	× 356	1.73	0.18	0.54	38	0.023	[40]
Bacillus sphaericus	230 (6 × 38,000) (needle-like crystal)	× 340	18.9	0.23	1.7	28	0.010	[41]
Bacillus cereus	255 (6 × 42,000)	—						[42]
Bacillus srearothermophilus (*Escherichia coli* clone cell)	235 (6 × 39,465)	× 30						[43]
Thermus termophilus	290 (6 × 48,000)	× 85	4.2	0.12	0.75	59	0.035	[44]
Streptomyces clavuligerus	92	× 38	9.1	0.5	1.1	20	0.14	[45]
Streptomyces phaeochromogenes	240 (6 × 39,000)	× 20	7.1	0.036	0.29	61	0.047	[46]
Streptomyces aureofaciens	395 (8 × 48,000)	× 714	5.0	0.11	0.56	6.7	0.029	[47]
Streptomyces fradiae	205–210 (4 × 51,000)	× 1180	10.0	0.18	0.23	12	0.050	[48]
Pseudomonas sp. (methylotrophic)	214 (4 × 53,000)	× 400						[49]
Desulfvibrio desuficans	—	× 56	—	—	5	24	0.055	[50]
Halobacterium cutirubrum	72.5 (monomer)	× 100						[51]
salinarium	60 (monomer)	× 500	K_m values is salt dependent					[52]
Anabaena cylindrica	270 (6 × 43,000)	× 700	0.4	0.014		8–133		[53]
Rhizobium lupini bacteroids	180 (4 × 41,000)		0.11					[54]

Table 6. Properties of purified Leucine dehydrogenases

Properties	B. sphaericus [55]	B. stearothermophilus [61]	B. cereus [60]	Clostridium thermoacticum [62]
Molecular weight	245,000	300,000	310,000	350,000
Subunit (M_r: 10^3)	6 (41)	6 (49)	8 (39)	6 (56)
Optimum pH				
deamination	10.7	11.0	11.5	—
amination	9.0–9.5	9.0–9.5	8.5–9.5	—
Coenzyme NAD: K_m (mM)	0.39	0.49	0.34	0.61
NADH: K_m (mM)	0.035	—	0.034	0.025
Substrate specificity (K_m: mM)				
(deamination): L-Leucine	100 (1.0)	100 (4.4)	100 (1.5)	100 (8.5)
L-Valine	74 (1.7)	98 (3.9)	61 (2.5)	93 (6.8)
L-Isoleucine	58 (1.8)	73 (1.4)	61 (1.0)	88 (3.9)
L-Norvaline	41 (3.5)	—	28 (2.9)	11 (5.6)
L-α-Aminobutyrate	14 (10)	—	24 (22)	19 (1.5)
L-Norleucine	10 (6.3)	—	6 (1.5)	15 (2.3)
γ-Methylallylglycine	8.2	—	—	—
tert-DL-Leucine	1.6	—	—	—
D-Leucine	0	0	0	0
(amination): α-Ketoisocaproate	100 (0.31)	100	100 (0.45)	100 (8.5)
α-Ketoisovalerate	126 (1.4)	167	154 (2.1)	130 (6.8)
α-Ketovalerate	76 (1.7)	86	51 (0.40)	67 (1.0)
α-Ketobutyrate	57 (7.7)	45	51 (1.5)	35 (7.3)
α-Ketocaproate	46 (7.0)	—	51 (1.2)	44 (6.2)
α-Keto-β-methylvalerate	—	—	104 (0.4)	—
α-Keto-γ-mercaptobutyrate	—	—	37 (2.1)	—

functions in the L-alanine catabolism and spore germination, in which the enzyme is used to supply the cells with pyruvate as an energy source and with ammonia as a nitrogen source. In cells of *Rhodopseudomonas capsulata* [37], *Streptomyces clavuligerus* [38], and *Anabeana cylindrica* [39], the enzyme participates in ammonia assimilation in the presence of high concentrations of ammonia. Ala DHs differ with respect to their subunit structures. The enzyme of halophilic bacteria is a monomer, the *Pseudomonas* sp. enzyme is a tetramer, *Bacillus* and *Thermus* enzymes are hexameric, and the *Streptomyces aureofaciens* enzyme is an octamer (Table 5). The enzyme requires NAD as a cofactor, which cannot be replaced by NADP. The substrate specificity of the enzyme for the oxidative deamination is high: L-alanine is exclusively deaminated. The specificity for keto acids is lower than that for amino acids: α-ketobutyrate, α-ketovalerate, and 3-hydroxypyruvate are aminated as well as pyruvate. The optimum pHs for the oxidative deamination and reductive amination are pH 10–10.5 and 8–9, respectively. The allosteric regulation of the enzyme by nucleotides and coenzymes has not been investigated.

2.3 Leucine Dehydrogenase

Leu DH (EC 1.4.1.9) catalyzes the reversible deamination of L-leucine and several other aliphatic amino acids to their keto analogs (Table 6). It occurs ubiquitously in *Bacillus* species [55] and functions catabolically in the bacterial metabolisms of L-branched chain amino acids [56]. It has been suggested that the enzyme plays an important role in spore germination in cooperation with Ala DH [57, 58]. Leu DH has been purified from *B. sphaericus* [55, 59], *B. cereus* [60], *B. stearothermophilus* [61], and *Clostridium thermoaceticum* [62]. The Leu DHs from *B. sphaericus*, *B. stearothermophilus*, and *C. thermoacetinum* are hexamers, and the *B. cereus* enzyme is an octamer with identical subunits [55, 62–64]. The substrate specificity of Leu DH is much lower than those of Glu DH and Ala DH. Ammonia is an exclusive amino donor; hydroxyamine, methylamine, glutamine, and asparagine are inert as the amino donor [55]. The low substrate specificity for keto acids is advantageous for the synthesis of enantiomers of various amino acids as described below. The thermostable enzyme from thermophiles such as *B. stearothermophilus* and *C. thermoaceticum* is more stable under various conditions than the enzyme from mesophiles [61, 62]. Thiol reagents such as *p*-chloromercuribenzoate and $HgCl_2$, and pyridoxal 5′-phosphate [65, 66] inhibit the enzyme.

2.4 Phenylalanine Dehydrogenase

Phe DH was first found in *Brevibacterium* sp. isolated from soil and can be used as a catalyst for the synthesis of L-phenylalanine, which is a starting material of an artificial sweetner, L-aspartyl-L-phenylalanine methyl ester (aspartame). It occurs in several gram-positive aerobic spore-forming bacteria and actinomycetes,

T. Ohshima and K. Soda

Table 7. Characteristics of phenylalanine dehydrogenases from various bacteria

Properties	S. urea	B. sphaericus	B. badius	R. sp.	R. maris	T. intermedius
Molecular weight (10^3)	305	340	310–360	—	70	270
Subunits (M_r; 10^3)	8 (38–39)	8 (41)	8 (41–42)	—	2 (36)	6 (45)
Optimum pH						
deamination	10.5	11.3	10.4	10.1	10.8	10.8
amination	9.0	10.3	9.4	9.25	9.8	9.2
Coenzyme NAD: K_m (mM)	0.14	0.17	0.15	0.27	0.25	—
NADH: K_m (mM)	0.072	0.025	0.21	0.13	0.043	—
Substrate specificity (K_m; mM)						
(deamination): L-Phenylalanine	100 (0.096)	100 (0.22)	100 (0.106)	100 (0.75)	100 (3.8)	100 (0.22)
L-Tyrosine	5.4	72 (0.50)	9.0	12 (3.1)	2.0	0
L-Tryptophan	5.0	1.2	4.0	2 (11)	7.5	0
L-Methionine	4.1	3.0	8.0	4 (0.43)	5.4	0
L-Ethionine	7.0	3.1	—	—	—	—
L-Leucine	2.3	1.3	3.0	—	2.0	3.9
L-Valine	3.1	1.4	4.0	—	0	0
D-Phenylalanine	0	0	0	0	0	0
(amination): Phenylpyruvate	100 (0.16)	100 (0.40)	100 (0.106)	100 (0.16)	100 (0.50)	100 (0.045)
p-Hydroxyphenylpyruvate	24	136 (0.34)	53	5 (2.4)	9 (1.3)	0
Indolpyruvate	0.73	0.39	—	3 (7.7)	5	—
α-Keto-4-methylmercapto-butyrate	27	11	16	33 (2.1)	9	0
α-Ketoisocaproate	13	7.3	13	—	1.2	5.5
Reference	(4)	(4)	(69)	(72)	(70)	(71)

S.: *Sporosarcina*, B.: *Bacillus*, R.: *Rhodococcus* and T.: *Thermoactinomyces*

and has been purified to homogeneity from several bacteria (Table 7). The production of bacterial Phe-DH is induced by the addition of L-phenylalanine and probably plays a role in the degradation of L-phenylalanine [67]. The enzymes from *Sporosarcina urea* [4, 68], *B. sphaericus* [68], and *B. badius* [69] are octamers, whereas dimeric and hexameric structures are seen in the *R. maris* [70] and *Thermoactinomyces intermedius* [71] enzymes, respectively. Like in other amino acid dehydrogenases a high activity is found at an extremely alkaline pH range. The substrate specificity for amino acids and keto acids differs markedly among the various enzyme sources. The *B. sphaericus* enzyme acts on L-tyrosine as well as L-phenylalanine, whereas the *T. intermedius* enzyme is highly specific for L--phenylalanine (Table 7).

2.5 Other Amino Acid Dehydrogenases

Among the amino acid dehydrogenases, *meso*-α,ε-diaminopimelate DH (EC 1.4.1.16) is unique because it acts on an amino group of the substrate with D-configuration to form L-α-amino-ε-ketopimelate [73, 74]. *meso*-α,ε-Diaminopimelate is a key intermediate in bacterial lysine synthesis and a constituent of the cell wall of certain bacteria. The enzyme functions in the formation of *meso*-α,ε-diaminopimelate from L-α-amino-ε-ketopimelate in bacterial lysine synthesis. The enzyme was purified from *B. sphaericus* [75], *Corynebacterium glutamicum* [76], and *Brevibacterium* sp. [77], and consists of two identical subunits with a molecular weight of 35,000 to 40,000. The *B. sphaericus* and *Brevibacterium* sp. enzymes require exclusively NADP as a coenzyme and the *C. glutamicum* enzyme requires either NADP or NAD. The enzyme acts exclusively on the *meso*-form of the substrate; neither the D- nor L-isomer serves as a substrate.

An NADP-dependent Val DH (EC 1.4.1.8) occurs is higher plants [78] and an NAD-dependent enzyme is found in some bacteria [79–81]. The Val DH is similar to the Leu DH. However, it is different from Leu DH in terms of its substrate specificity: L-valine is the most preferred substrate and the *Aerobacter aerogenes* enzyme was found to be immunologically different from the *B. stearothermophilus* Leu DH [80]. Lys DH (EC 1.4.1.15) was recently purified from *Agrobacterium tumefaciences* and some of the fundamental properties were reported [82]. Try DH was found in higher plants [83], although they have not been studied in detail.

3 Stereospecificity of Hydrogen Transfer of Coenzyme and Kinetic Mechanism of Amino Acid Dehydrogenase Reaction

NAD(P)-dependent dehydrogenases show either pro-*R* (A-) or pro-*S* (B-) stereospecificity for hydrogen transfer from the C-4 position of the nicotinamide moiety of NAD(P)H to the substrate (Fig. 1) [84]. The stereospecificity of the hydrogen transfer has been investigated by means of NAD(P)H labeled stereospecifically with ^{2}H or ^{3}H at the C-4 position of the nicotinamide ring as described below. The

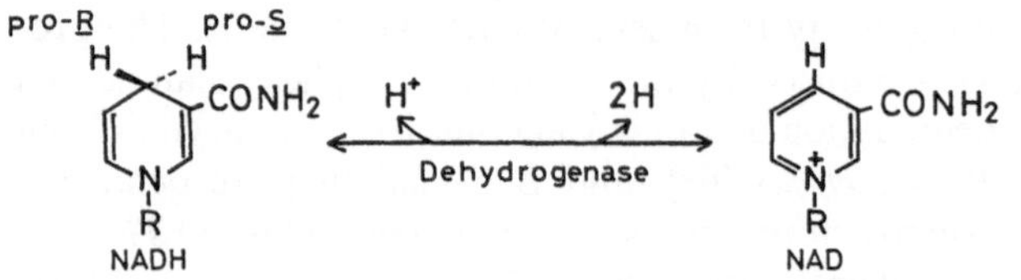

Fig. 1. Stereospecificity of hydrogen transfer of NADH catalyzed with dehydrogenases.
R: ADP-ribosyl

Table 8. Stereospecificity of hydrogen transfer of coenzymes

Enzyme		Enzyme source	Stereospecificity	Ref.
Glu DH:	NAD	Yeast	Pro-S specific	[85]
	NAD(P)	Bovine liver	Pro-S specific	[85]
	NADP	Plant	Pro-S specific	[85]
Ala DH:	NAD	B. subtilis	Pro-R specific	[86]
	NAD	B. sphaericus	Pro-R specific	[41]
Leu DH:	NAD	B. sphericus	Pro-S specific	[55]
Phe DH:	NAD	B. sphaericus	Pro-S specific	[64]
	NAD	T. intermedius	Pro-S specific	[71]
Lys DH:	NAD	Agrobacterium tumefaciens	Pro-R specific	[87]
Val DH:	NAD	Streptococcus thermovulgaris	Pro-S specific	[88]

stereospecificity of the amino acid dehydrogenases which have so far been investigated is listed in Table 8. Amino acid dehydrogenases except Ala DH and Lys DH show pro S-stereospecificity. We found that Leu DH and Ala DH of *B. sphaericus* show the opposite stereospecificity for the reaction of α-ketobutyrate to L-α-aminobutyrate [55]. This shows that the stereospecificity is an inherent characteristic of individual dehydrogenase and is independent of the enzyme reaction and source. The physiological significance of the stereospecificity of the hydrogen transfer is obscure.

A series of steady-state kinetic analyses provides information about the reaction mechanism [89]. All of the reactions catalyzed by amino acid dehydrogenases proceed via the formation of a ternary complex with sequential mechanism and not with ping-pong mechanism. However, diversity is found in the manner of substrate binding and product release. The reaction of bovine liver NAD(P)-Glu DH proceeds via a random ordered binding (Fig. 2) [7]. Microbial and plant Glu DHs catalyze the reaction through sequential (compulsory) ordered binary-ternary mechanism [7, 90]. In the NAD-Glu DH reaction, NAD binds first to the enzyme followed by glutamate, and the products are released in the order of ammonia, α-ketoglutarate, and NADH. The reverse order of the product release (ammonia and α-ketoglutarate) is found for NADP-Glu DH [91]. The *Bacillus* Ala DH reaction also proceeds through different kinetic mechanisms. The enzymes from the mesophilic *B. sphaericus* [41], *B. subtilis* [92, 93] strains and from the thermophilic *B. sphaericus* [94] differ from each other in the way

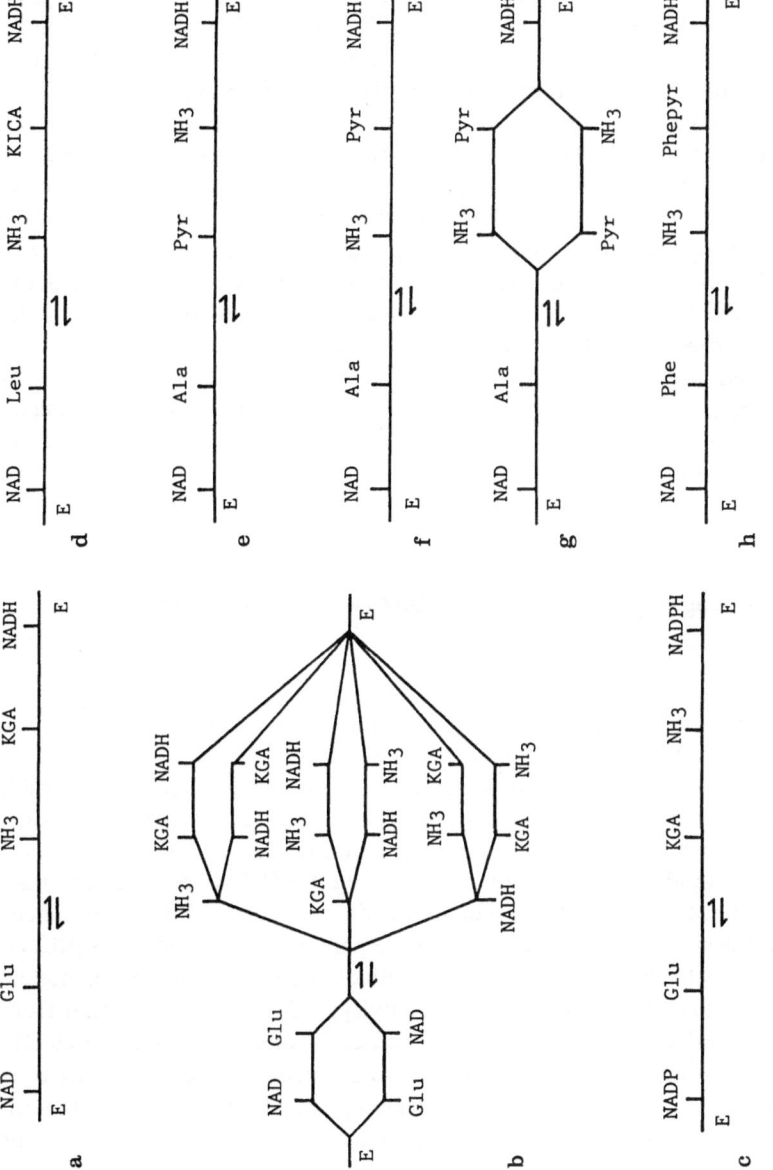

Fig. 2a–h. Kinetic mechanisms of amino acid dehydrogenases. **a** NAD-Glu DH, **b** NAD(P)-Glu DH, **c** NADP-Glu DH, **d** Leu DH, **e** Ala DH of mesophilic *B. sphaericus*, **f** Ala DH of *B. subtilis*, **g** Ala DH of thermophilic *B. sphaericus*, and **h** Phe DH. KGA: α-ketoglutarate, KICA: α-ketoisocaproate, Pyr: pyruvate, and Phepyr: phenylpyruvate

in which ammonia and pyruvate are released (Fig. 2). It is intersting that amino acid dehydrogenases from different enzyme sources differ in their kinetic mechanism, although the significance of the diversity in the catalytic and physiological functions of the enzymes is unknown. From kinetic analysis of the thermophilic *B. sphaericus* Ala DH, the presence of an abortive tenary dead-end inhibition by the formation of an enzyme-NAD-pyruvate complex was suggested [94]. This abortive dead-end inhibition may play an important role in preventing the enzyme functioning in L-alanine synthesis.

In the bovine liver Glu DH reaction, the formation of α-iminoglutarate as an intermediate has been proposed [7, 95]. Incubation of Glu DH with α-ketoglutarate and ammonia leads to the formation of α-iminoglutarate, which is trapped by borohydride reduction to yield L-glutamate. In the reaction scheme involving α-iminoglutarate, the ε-amino group of the lysine residue at the active site is considered to be involved in the enzyme-substrate (α-ketoglutarate) complex [7]. The formation of the α-amino intermediate strengthens the case for a random ordered-binding mechanism. The intermediate stages of the reactions catalyzed by NAD and NADP-Glu DHs and other amino acid dehydrogenases have not yet been clarified because the enzyme reactions proceed through the sequential ordered-binding mechanism.

4 Gene Cloning and Primary Structure of Amino Acid Dehydrogenases

In recent years, the thermostable Leu DH gene of *B. stearothermophilus* was cloned into *E. coli* [96, 97]. Chromosomal DNA fragments of *B. steraothermophilus* obtained by *Sal* I digestion were ligated into the *Sal* I site of plasmid, pBR322. Recombinant *E. coli* cells producing Leu DH were screened by means of a replica printing method involving a color reaction for Leu DH. This method is favorable for rapidly screening clones producing NAD(P)-dependent enzymes. Of about 2700 recombinants, two clones producing Leu DH were obtained. They carried plasmids of 8.95 and 6.0 kilo base pair (kbp), designated as pICD1 and pICD2, respectively. Cell extracts from both clones showed an approximately fifty-fold higher Leu DH activity (specific activity: 3.5 units/mg) than that from *B. stearothermophilus* (0.75 units/mg). In *E. coli* cells containing pICD1, Leu DH constitutes approximately 3% of the total soluble protein; the total activity is about 13-times as high as that of *B. stearothermophilus*. Upon heating of the cell extract of the *E. coli* strain containing pICD1 at 70 °C under acidic conditions (pH 5.4) for 30 min, the specific activity is increased markedly without a significant loss of activity and with a yield of about 93%. Other proteins produced by *E. coli* cells were denatured under these conditions. The enzyme was purified to homogeneity by two further steps, DEAE-Toyopearl 650 M and Sephadex G-200 column chromatography with a yield of approximately 70%. Approximately 1 g of pure Leu DH was obtained from 1 kg (wet weight) of *E. coli* cells within 5 days. The enzyme is now commercially available (Unitika Ltd., Osaka, Japan).

The cloned Leu DH gene plus flanking regions (2.26 kbp) facilitated the sequencing of the 1287 bp coding region. The entire amino acid sequence (429 residues) was determined from the DNA sequence and compared with that obtained by partial peptide sequencing [97].

The structural gene of the thermostable Ala DH of *B. stearothermophilus* was cloned in *E. coli* cells in a similar manner; the amount of Ala DH corresponded to approximately 4% of the total soluble protein in the cell extract, and its specific activity (2.37 units/mg) was about 60-fold higher than that from *B. stearothermophilus* [98]. The thermostable Ala DH was also purified to homogeneity with an yield of approximately 47% by heat treatment of the cell extract (70 °C, 1 h, followed by three chromatographic steps. The homogeneous enzyme is commercially supplied (Unitika Ltd., Osaka, Japan). In addition, the DNA sequence of the enzyme was determined in a manner similar to that of Leu DH [43].

The amino acid sequence of various amino acid dehydrogenases has been determined by peptide and DNA sequencing methods (Table 9). Although a

Table 9. Amino acid sequences of amino acid dehydrogenaes

Enzyme		Enzyme source	Number of amino acids sequenced	Ref.
Glu DH:	NAD(P)	Human	505	[99]
	NAD(P)	Bovine liver	500	[7]
	NAD(P)	Chicken liver	503	[7]
	NAD(P)	Rat liver	500	[7]
	NAD	*Neurospora crassa*	669	[100]
	NADP	*N. crassa*	452[a]	[101]
	NADP	*Saccharomyces cerevisea*	454[a]	[102]
	NADP	*E. coli*	447[a]	[103]
Ala DH:	NAD	*B. sphaericus*	372[a]	[43]
	NAD	*B. stearothermophilus*	372[a]	[43]
Leu DH:	NAD	*B. stearothermophilus*	429[a]	[97]
Phe DH:	NAD	*B. sphaericus*	381[a]	[104]
meso-Diamino-pimerate DH:	NADP	*Corynebacterium grutamicum*	320[a]	[105]

[a] Amino acid sequence was determined by nucleotide sequencing

computer-aided search of the National Biochemical Research Foundation protein sequence data bank [106] revealed rather low over all sequence similarities among amico acid dehydrogenases, a partial sequence of about 30 residues in the nicotinamide coenzyme binding region is common in all enzymes [97, 98]. The coenzyme binding domain which binds the adenine nucleotide moiety shows a high degree of conservation of tertiary structures; it consists of a two-stranded parallel β-sheet and one α-helix with virtually identical arrangement. Further decipherment of partial sequences of the domains and other important residues which participate in the substrate binding, and studies of the reaction mechanism and of the stereospecificity of hydrogen transfer are now under investigation. The availability of the cloned gene and of the corresponding sequence will allow further structural studies of amino acid dehydrogenases by site directed mutagenesis.

5 Applications of Amino Acid Dehydrogenases

5.1 Production of L- and D-Amino Acids

In recent years, enzyme and whole-cell bioreactor methods as well as cultivation and chemical methods have been developed for L-amino acid production [107–109]. Continuous production of L-leucine and other aliphatic L-amino acids from their corresponding keto analogs and ammonium formate was studied with an ultrafiltration membrane reactor (molecular cut-off at M_r 5000) containing Leu DH, yeast formate dehydrogenase (FDH, M_r 80,000) and NADH or NAD bound covalently to polyethylene glycol (PEG, M_r 20,000) [110]. PEG-NADH cannot penetrate through the membrane while FDH catalyzes regeneration of the PEG-NADH with formate. It is considerably stable and cheaply available, and its reaction is irreversible. α-Ketoisocaproate and ammonium formate are continuously pumped into the reactor, and L-leucine is produced together with CO_2 (Fig. 3). The enzyme membrane reactor system is applicable to the production of several other aliphatic L-amino acids containing L-methionine and L-*tert*-leucine due to the broad substrate specificity of Leu DH [111, 112]. The efficiency of this reactor system depends mainly on the stability and abundant supply of en-

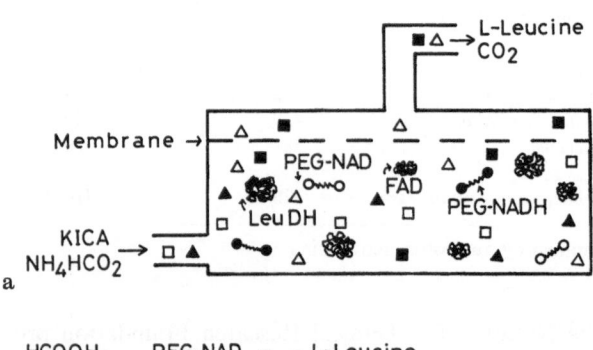

Fig. 3a–b. Reaction sytem for continuous production of L-leucine with coenzyme regeneration. **a** Enzyme membrane reactor and **b** the enzyme reaction system for L-leucine production. The substrate solution containing α-ketoisocaproate (KICA) and ammonium formate is pumped into the reactor at a constant flow rate

zymes. The thermostable Leu DH produced abundantly by *E. coli* cells has a long half-life, and thus is more useful in the reactor system [113]. A similar enzyme membrane reactor system with Ala DH has been developed for the production of L-alanine [114] and L-3-fluoroalanine [115]. Phe DH is used for the continuous production of L-phenylalanine, which is important as a starting material for an artifical sweetener, aspartame [116, 117].

The multi-enzyme system consisting of Ala DH, FDH, alanine racemase, and D-amino acid aminotransferase has been developed for the production of D-amino acids, which are useful for the chemical synthesis of β-lactam antibiotics and bioactive peptides (Fig. 4) [118]. This procedure is based on the strict stereo-specificity and low structural specificity for substrates of D-amino acid aminotransferase and the very high substrate specificity of alanine racemase. Ala DH and FDH are used for L-alanine production and NADH regeneration, respectively.

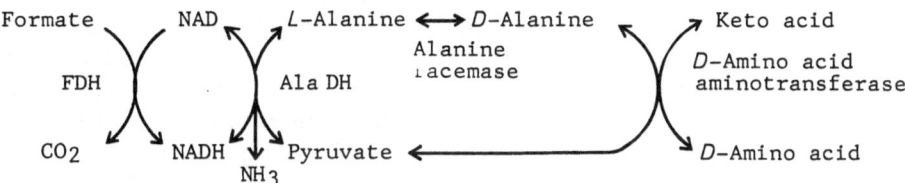

Fig. 4. Principle of multi-enzyme system for the synthesis of acids. The corresponding α-keto acid and ammonium formate are used as the substrates. A small amount of NAD, L-alanine, and pyridoxal 5′-phosphate are added to the reaction mixture for regeneration of D-alanine

Various D-amino acids such as D-glutamate, D-methionine, and D-valine are produced. A similar system consisting of Glu DH and glutamate racemase is applicable to the effective production of D-alanine, D-aspartate, D-α-amino-butyrate, and D-valine [119].

The continuous method for L-leucine production by means of a gel column in which Leu DH and FDH are coimmobilized was developed [120]. In addition, a continuous production of L-alanine with immobilized cells of *Clostridium butyricum* containing Ala DH was studied under high pressure of hydrogen [121]. Amino acid dehydrogenases are used for the preparation of ^{15}N-labeled amino acids such as L-^{15}N-alanine [122, 123], L-^{15}N-glutamate [124], and L-^{15}N-leucine [125] as well.

5.2 Preparation of Stereoselectively Deuterated NADH and NADPH

NADH and NADPH deuterated stereospecifically at C-4 of the nicotinamide moiety are useful for the determination of the stereospecificity of NAD(P)-dependent dehydrogenases and for the study of deuterium isotope effect in the reactions catalyzed by NAD(P)-dependent enzymes. We have developed new methods for the preparation of [$4R$-^{2}H] and [$4S$-^{2}H] NAD(P)H by coupling amino acid dehydrogenase and amino acid racemase reactions [126]. Amino acid racemases inherently catalyze the exchange of the α-H of the substrate amino acids with deuterium during racemization in ^{2}H$_2$O [127]. [$4R$-^{2}H]NADH is produced from D-alanine and NAD by coupling of the reactions of alanine racemase and Ala DH in ^{2}H$_2$O (Fig. 5). D-Alanine is racemized to form D- and L-[α-^{2}H]-alanine by alanine racemase. The L-isomer is deaminated to pyruvate by Ala DH (pro-R stereospecific hydrogen transfer), and the α-^{2}H is transferred to the pro-R position at C-4 of the nicotinamide moiety of NADH. The yield of purified NAD^{2}H was 72%. [$4S$-^{2}H]NADH is produced in the same manner as described above with amino acid racemase with low substrate specificity [128] and Leu DH (pro-S stereospecific). [$4S$-^{2}H]NADPH is also produced in a similar way by means of glutamate racemase and NADP-Glu DH (pro-S stereospecific) [129]. In addition, we established a simple method for determining the stereospecificity of hydrogen transfer of NAD(P)H by an NAD(P)-dependent dehydrogenase. After incubation of the reaction mixture containing alanine racemase, Ala DH, D-alanine, an NAD-dependent dehydrogenase whose stereospecificity is to be examined, and the oxidized form of substrate in ^{2}H$_2$O, the C-4 proton of NAD is determined by ^{1}H NMR. If the stereospecificity of hydrogen transfer of a dehydrogenase is the same as that of Ala DH, the C-4 hydrogen of NAD is fully retained and a doublet specific for it appears in the ^{1}H-NMR spectra, and the final product is deuterated. In the combination of dehydrogenases having different stereospecificities, the C-4 hydrogen of [$4R$-^{2}H]NADH is transferred to the substrate, and the C-4 deuterium remains in NAD; consequently, no peak specific for the C-4 hydrogen of NAD appears in the ^{1}H NMR spectra.

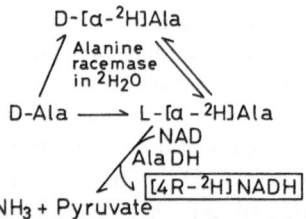

Fig. 5. Principle for [$4R$-^{2}H] NADH preparation by coupling of alanine recemase and alanine dehydrogenase

The stereospecificity of hydrogen transfer by a dehydrogenase is determined in the same manner with the alternative enzyme system. Leu DH and amino acid racemase with low substrate specificity are substituted for Ala DH and alanine racemase, respectively in the reaction system. In addition, the stereospecificity of hydrogen transfer of NADPH is determined in a similar manner with glutamate racemase and NADP-Glu DH (pro-S stereospecific) [129]. Thus, we can readily determine the stereospecificity by ^{1}H-NMR measurement without isolation of the coenzyme and products; this is the most rapid and convenient method currently available for this purpose.

5.3 Application of Amino Acid Dehydrogenase to Specific Quantitative Analyses

Enzymatic analysis of amino acids, keto acids and ammonia is an important tool in clinical chemistry, bioprocess control, and nutrition studies. It is more simple and inexpensive compared with ion-exchange high performance liquid chromatography methods. The methods are based on the increase and decrease in concentration of NADH in the enzyme reaction. The spectrophotometric analyses for the determination of various amino acids and keto acids with amino

Table 10. Enzymatic determination using amino acid dehydrogenases

Amino acid dehydrogenase	Amino acids, keto acids, enzymes
Glu DH	L-glutamate, α-ketoglutarate, ammonia, urea (urease), D-glutamate (glutamate racemase), L-glutamine (glutamianse), L- citrulline (citrulline hydrolase), L-proline, glutamate-oxaloacetate aminotransferase, branched chain amino acid-α-ketoglutarate aminotransferase.
Ala DH	L-alanine, pyruvate, alanine aminotransferase, γ-glutamyl cyclotransferase, amino acylase.
Leu DH	L-leucine, L-valine, L-isoleucine, α-ketoisocaproate, α-ketoisovalerate, α-keto-β-methylvalerate, aminopeptidase, D-amino acid aminotransferase, L-methionine (methionine γ-lyase).
Diaminopimerate DH	meso-α, ε-diaminopimerate
Phe DH	L-phenylalanine, L-tyrosine, phenylpyruvate.

( ): Enzyme used in the conjugated reaction

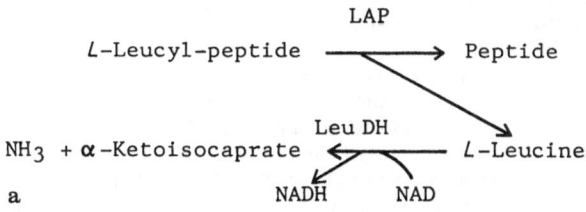

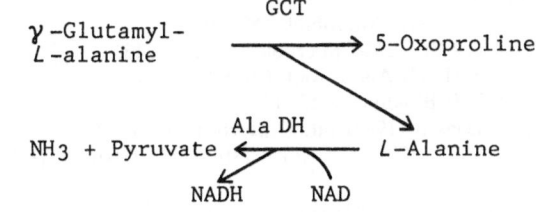

Fig. 6a, b. Principles of serum aminopeptidase (LAP) and γ-glutamyl cyclotransferase (GCT) assays. **a** LAP and **b** GCT assay systems

acid dehydrogenases have been developed as listed in Table 10 [130]. The methods are applicable to the assay of enzymes which are marker enzymes of human diseases. Leu DH is used for the assay of serum leucine aminopeptidases, which is related to liver diseases [131, 132], and Ala DH is for erythrocytic γ-glutamyl cyclotransferase, which is a marker of malignant hematopoietic disease (Fig. 6) [133]. Phe DH of *Thermoactinomyces intermedius* is used for the specific determination of L-phenylalanine and phenylpyruvate, and is therefore applicable to diagnosis of neonatal hyperphenylalaninemia and phenylketonuria [134].

In general, the K_m values of amino acid dehydrogenases for ammonia are very high. For *B. sphaericus* Leu DH, Ala DH and Phe DH the values are 220, 28.2, and 78 mM, respectively. In contrast the NADP-Glu DH of *Proteus* sp. exhibits a relatively low K_m value (1.1 mM) for ammonia. Thus, the Glu DH is applied to the determination of ammonia and urea by coupling it with the urease reaction [135]. Thermostable amino acid dehydrogenases are advantageous for these analyses because of their high stability and abundant supply by the cloned cells as mentioned above.

6 References

1. Engel PC, Dalziel K (1967) Biochem. J. 105: 691
2. Yoshisa A, Freese E (1965) Biochem. Biophys. Acta 96: 248
3. Sanwal BD, Zink MW (1961) Arch. Biochem. 94: 430
4. Asano Y, Nakazawa A, Endo K (1987) J. Biol. Chem. 262: 10346
5. Sanwal BD, Lata M (1961) Can. J. Microbiol. 7: 319
6. Sanwal BD, Lata M (1961) Nature (London) 190: 286
7. Smith EL, Austen BM, Blumenthal KL, Nyc JF (1975) In: Boyer PD (ed) The Enzymes 3rd edn. Academic Press, New York (vol 11A) p 293
8. Lehman F, Pfleiderer G (1968) Hoppe-Seiler's Z. Phydsol. Chem. 350: 609
9. Teller J K (1988) Insect Biochem. 18: 101
10. Itagaki T, Ian B, Wiskich J T (1988) Phytochemistry 27: 3373
11. Botton B, Msatef Y (1983) Physiol. Plant. 59: 438
12. Javede Q, Merrett M J (1986) New Phytol. 104: 407
13. Sofin AV, Shatilov VR, Kretovich VL (1983) Biokimiya (Moscow) 48: 2056
14. Stevens L, Duncan D, Robertson P (1989) FEMS Microbiol. Lett. 57: 173
15. Uno I, Matsumoto K, Adachi K, Ishikawa T (1984) J. Biol. Chem. 259: 1288
16. Martinez M, Martinez A, Urkijo I, Lama MJ, Serra JL (1988) J. Bacteriol. 170: 4897
17. Weining S, Nicholas DJD (1987) Phytochemistry 26: 2151
18. Saito H, Yamamoto I, Ishimoto M (1988) J. Gen. Appl. Microbiol. 34: 377
19. Kimura K, Miyakawa A, Imai T, Sasakawa T (1977) J. Biochem. 81: 467
20. Montsala P (1985) Biochem. Int. 10: 955
21. Hammer B A, Johnson E A (1988) Arch. Microbiol. 150: 460
22. Duchars M G, Attwood M M (1987) FEMS Microbiol. Lett. 38: 133
23. Misono H, Goto N, Nagasaki S (1985) Agric. Biol. Chem. 49: 117
24. Sokolov AP, Trotsenko UA (1987) Biokhimiya 52: 1417
25. Bonete MJ, Camacho ML, Cadenas E (1987) Int. J. Biochem. 19: 1149
26. Ogawa S, Rottenberg H, Brown TR, Shulman RG, Castillo RE, Glynn P (1978) Proc. Natl. Acad. Sci. USA 75: 1796
27. Josephs R, Borisky G (1972) J. Mol. Biol. 65: 127
28. Gore MG (1981) Int. J. Biochem. 13: 879
29. Veronese FM, Nyac JF, Degani Y, Brown DM, Smith EL (1974) J. Biol. Chem. 249: 7922

30. Hemmings BA (1980) J. Biol. Chem. 255: 7925
31. Mazon MJ, Hemmings BA (1979) J. Bacteriol. 139: 686
32. Wiggert BO, Cohen PP (1965) J. Biol. Chem. 240: 4790
33. Hooper AB, Hansen J, Bell R (1967) J. Biol. Chem. 242: 288
34. Struck J, Sizer IW (1960) Arch. Biochem. Biohys. 86: 260
35. Berberich R, Kaback M, Freese E (1968) J. Biol. Chem. 243: 1006
36. Ohshima T, Wandrey C, Sugiura M, Soda K (1985) Biotechnol. Lett. 7: 871
37. Johansson BC, Gest H (1976) J. Bacteriol. 128: 683
38. Aharonowitz Y, Freiddrich CG (1980) Arch. Microbiol. 125: 137
39. Rowell P, Stewart WDP (1976) Arch. Microbiol. 107: 115
40. Yoshida A, Freese E (1970) Mthods Enzymol. 17: 176
41. Ohshima T, Soda K (1979) Eur. J. Biochem. 100: 29
42. Porumb H, Vancea D, Muresan L, Presecan E, Lascu I, Petrescu I, Porumb T, Pop R, Barzu O (1987) J. Biol. Chem. 262: 4610
43. Kuroda S, Tanizawa K, Sakamoto Y, Tanaka H, Soda K (1990) Biochemistry 29: 1009
44. Vali Z, Kilar F, Lalatos S, Venyaminov SA, Zavodoszky P (1980) Biochim. Biophys. Acta 615: 34
45. Itoh N, Morikawa R (1983) Agric. Biol. Chem. 47: 2511
46. Aharonowitz Y, Friedrich CG (1980) Arch. Microbiol. 125: 137
47. Vancurova I, Vancura A, Volc J, Neuzil J, Flieger M, Basarova G, Behal V (1988) Arch. Microbiol. 150: 438
48. Vancura A, Vanculova I, Volc J, Jones SKT, Flieger M, Basarova G, Behal V (1989) Eur. J. Biochem. 179: 22
49. Bellion E, Tan F (1987) Biochem. J. 244: 565
50. Germano GJ, Anderson KE (1968) J. Bacteriol. 96: 55
51. Kim EK, Fitt PS (1977) Biochem. J. 161: 313
52. Keradjopoulos D, Holldorf AW (1979) Biochim. Biophys. Acta 570: 1
53. Rowell P, Stewart WDP (1976) Arch. Microbiol. 107: 115
54. Kazakova OV, Ivanushkin AG, Tsuprun VL, Kaftanova AS, Pushkin AV, Kretovich V (1988) Biokhimiya (Moscow) 53: 1864
55. Ohshima T, Misono H, Soda K (1978) J. Biol. Chem. 253: 5719
56. Zink MW, Sanwal BD (1962) Arch. Biochem. Biophys. 99: 72
57. Hermier J, Lebeault JM, Zevako C (1970) Bull. Soc. Chim. Biol. 52: 1089
58. Hermier J, Rosseau M, Zevako C (1970) Ann. Inst. Pasteur Paris 118: 611
59. Soda K, Misono M, Mori K, Sakato H (1971) Biochem. Biophys. Res. Commun. 44: 931
60. Schütte H, Hummel W, Tsai H, Kula MR (1985) Appl. Microbiol. Biotechnol. 22: 306
61. Ohshima T, Nagata S, Soda K (1985) Arch. Microbiol. 141: 407
62. Shimoi H, Nagata S, Esaki N, Tanaka H, Soda K (1987) Agric. Biol. Chem. 51: 3375
63. Hiragi Y, Soda K, Ohshima T (1982) Makromol. Chem. 183: 745
64. Lunsdorf H, Tsai H (1985) FEBS Lett. 193: 261
65. Ohshima T, Yamamoto T, Misono H, Soda K (1978) Agric. Biol. Chem. 42: 1739
66. Ohshima T, Soda K (1984) Agric. Biol. Chem. 48: 349
67. Hummel W, Weiss N, Kula MR (1984) Arch. Microbiol. 137: 349
68. Asano Y, Nakazawa A (1985) Agric. Biol. Chem. 49: 3631
69. Asano Y, Nakazawa A, Endo K, Hibino Y, Ohmori M, Numao N, Kondo K (1978) Eur. J. Biochem. 168: 153
70. Misono H, Yonezawa J, Nagata S, Nagasaki S (1989) J. Bacteriol. 171: 30
71. Ohshima T, Sugimoto H, Soda K (1987) Abstr. Annu. Agric. Chem. Soc. Jpn. p 627
72. Hummel W. Schütte H, Schmidt E, Wandrey C, Kula MR (1987) Appl. Microbiol. Biotechnol. 26: 409
73. Misono H, Tamamoto T, Soda K (1976) Biochem. Biophys. Res. Commun. 72: 89
74. Misono H, Togawa H, Yamamoto T, Soda K (1979) J. Bacteriol. 137: 22
75. Misono H, Soda K (1980) J. Biol. Chem. 255: 10599
76. Ishino S, Yamaguchi K, Shirahata K, Arai K (1984) Agric. Biol. Chem. 48: 2557
77. Misono H, Ogasawara M, Nagasaki S (1986) Agric. Biol. Chem. 50: 1329
78. Kagan ZS, Polykov VA, Kretovich VL (1968) Biokhimiya 33: 89

208   T. Ohshima and K. Soda

79. Omura S, Tanaka Y, Mamada H, Masuma R (1983) J. Antibiot. 36: 1792
80. Ohshima T, Soda K (1987) Vitamins (Japanese) 61: 299
81. Vancurova I, Vancura A, Volc J, Neuzil J, Flieger M, Basarova G, Behal V (1988) J. Bacteriol. 170: 5192
82. Misono H, Uehigashi H, Morimoto E, Nagasaki S (1985) Agric. Biol. Chem. 49: 2253
83. Ebeid MM, Dimova S, Kutacek (1985) Biologia Plantarum (Praha) 27: 413
84. Popjak G (1970) In: Boyer PD (ed) The Enzymes 3rd edn. Academic Press, New York (vol 2) p 116
85. You K, Arnold LJ Jr, Allison WS, Kaplan NO (1978) Trends Biochem. 3: 265
86. Alizade MA, Bressler R, Brendel K (1975) Biochim. Biophys. Acta 397: 5
87. Hashimoto H, Misono H, Nagata S, Nagasaki S (1989) Agric. Biol. Chem. 53: 1175
88. Ohshima T, Wada, Higashitani S, Soda K (1988) Vitamins (Japanese) 62: 217
89. Cleland WW (1970) In: Boyer PD (ed) The Enzymes 3rd edn. Academic Press, New York (vol 2) p 1
90. LéJohn HB, Jackson SG, Krassen GR, Sawula RV (1969) J. Biol. Chem. 244: 5346
91. LéJohn HB, Suzuki I, Wright JA (1968) J. Biol. Chem. 243: 118
92. Grimshaw CE, Cleland WW (1981) Biochemistry 20: 5650
93. Grimshaw CE, Cook PF, Cleland WW (1981) Biochemistry 20: 5650
94. Ohshima T, Sakane M, Yamazaki T, Soda K (1989) Abstr. Annu. Agric. Chem. Jpn, p 48 (Eur. J. Biochem. in press)
95. Sirinivasan R, Fischer HF (1985) Biochemistry 24: 618
96. Ohshima T, Soda K (1985) Fermentation and Industry (Hakko to Taisha, in Japanese) 43: 919
97. Nagata S, Tanizawa K, Esaki N, Sakamoto Y, Oshima T, Tanaka H, Soda K (1989) Biochemistry 27: 9056
98. Soda K, Nagata S, Tanaka H, Ohshima T, Sakamoto Y (1985) Japanese Patents 60 180580 and 60 180590 (J. Ferment. Technol. 69: 154)
99. Amuro N, Yamaura M, Goto Y, Okazaki T (1988) Biochem. Biophys. Res. Commun. 152: 1935
100. Harberland ME, Smith EL (1980) J. Biol. Chem. 255: 7984
101. Kinnaird JH, Fincham JRS (1983) Gene 26: 253
102. Moye WS, Amuro N, Mohana Rao JK, Zalkin H (1985) J. Biol. Chem. 260: 8502
103. McPherson MJ, Wootton JC (1983) Nucleic Acid Res. 11: 5257
104. Okazaki N, Hibino Y, Asano Y, Ohmori M, Numao N, Kondo K (1988) Gene 63: 337
105. Ishino S, Mizukami K, Yamaguchi K, Katsumata R, Araki K (1987) Nucleic Acids Res. 15: 3917
106. George DG, Barker WC, Hunt LT (1986) Nucletic Acids Res 14: 11
107. Soda K, Tanaka H, Esaki N (1983) In: Dellweg H (ed) Biotechnology Verlag Chemie International (vol 3) p 479
108. Hamilton KB, Hisao HY, Swann WE, Anderson DM, Delente JJ (1985) Trends Biotechnology 3: 64
109. Chibata I, Tosa T, Sato T (1987) In: Kennedy JF (ed) Biotechnology VCH (vol 7a) p 653
110. Buckmann AF, Morr M, Kula MR (1987) Biotechnol Appl. Biochem 9: 258
111. Wichmann R, Wandrey C, Buckmann AF, Kula MR (1981) Biotechnol. Bioeng. 23: 2789
112. Kula MR, Wandrey C (1988) Method Enzymol 136: 34
113. Ohshima T, Wandrey C, Kula MR, Soda K (1985) Biotechnol. Bioeng. 27: 1616
114. Fiolitaktis E, Wandrey C (1983) In: Enzyme Technology Proc., Rotenburg Fermentation Symposium 1982, Springer-Verlag, p 272
115. Ohshima T, Wandrey C, Conrad D (1989) Biotechnol. Bioeng. 34: 394
116. Hummel W, Schütte H, Schmidt E, Wandrey C, Kula MR (1987) Appl. Microbiol. Biotechnol. 26: 409
117. Schmidt E, Fiolitaktis E, Wandrey C (1987) In: Laskin AI, Mosbach K, Thomas D, Wingard LD (eds) Enzyme Engineering 8, Ann, N.Y. Acid Sci. 501: 434
118. Soda K, Tanaka H, Tanisawa K, Esaki N (1988) In: Durand G, Bobichon L, Florent J (eds) 8th International Biotechnology Symposium, Paris (Vol 1) p 361
119. Nakajima N, Tanizawa K, Tanaka H, Soda K (1988) J. Biotechnol. 8: 243

120. Kajiwara S, Maeda H (1987) Agric. Biol. Chem. 51: 2879
121. Matsunaga T, Matsunaga M, Nishimura S (1985)
121. Matsunaga T, Matsunaga M, Nishimura S (1985) Biotechnol. Bioeng. 27: 1277
122. Moscanu A, Niac G, Ivanof A, Gorun V, Palibroda N, Vargha E, Bologa M, Barzu O (1982) FEBS Lett. 143: 153
123. Presecan E, Ivanof A, Mocanu A, Palibroda N, Bologa M, Gorun V, Oarga M, Barzu O (1987) Enzyme Microb. Technol. 9: 663
124. Bojan D, Bologa M, Niac G, Vargha E, Barzu O (1980) Anal. Biochem. 101: 23
125. Wandrey C (1984) Forum Mikrobiologie (Sonderheft Biotechnologie) 7: 33
126. Esaki N, Shimoi H, Nakajima N, Ohshima T, Tanaka H, Soda K, J. Biol. Chem. 264: 9750
127. Shen S, Floss HG, Kumagai H, Yamada H, Esaki N, Soda·K, Wasserman SA, Walsh C (1983) J. Chem. Soc. Chem. Commun. 82
128. Inagaki K, Tanizawa K, Tanaka H, Soda K (1987) Agric. Biol. Chem. 51: 173
129. Nakajima N, Tanizawa K, Tanaka H, Soda K (1986) Agric. Biol. Chem. 50: 2823
130. Bergmeiyer HU (ed) (1985) Methods of Enzymatic Analysis, 3rd edn (vol 8)
131. Takamiya S, Ohshima T, Tanizawa K, Soda K (1983) Anal. Biochem. 130: 266
132. Takamiya S, Ohshima T, Tanizawa K, Soda K (1983) Agric. Biol. Chem. 47: 893
133. Takahashi T, Kondo T, Ohno H, Minato S, Ohshima T, Mikuni S, Soda K, Taniguchi N (1987) Biochem. Med. Metabol. Biol. 38: 311
134. Ohshima T, Sugimoto H, Soda K (1988) Anal. Lett. 21: 2205
135. Murachi T, Tabata K (1986) Methods Enzymol. 137D: 260

Author Index Volumes 1–42